# WHERE

# WE MEET

# THE WORLD

**Also by Ashley Ward:**

*The Social Lives of Animals*

# WHERE
# WE MEET
# THE WORLD

## THE STORY OF
## THE SENSES

Ashley Ward

BASIC BOOKS

NEW YORK

Basic Books
Hachette Book Group
1290 Avenue of the Americas, New York, NY 10104
www.basicbooks.com

Printed in the United States of America

Originally published in 2023 by Profile Books Ltd. in Great Britain
First US Edition: March 2023

Published by Basic Books, an imprint of Perseus Books, LLC, a subsidiary of
Hachette Book Group, Inc. The Basic Books name and logo is a trademark
of the Hachette Book Group.

The Hachette Speakers Bureau provides a wide range of authors for speaking
events. To find out more, go to hachettespeakersbureau.com or email
HachetteSpeakers@hbgusa.com.

Basic books may be purchased in bulk for business, educational, or promotional use.
For information, please contact your local bookseller or Hachette Book Group Special
Markets Department at special.markets@hbgusa.com.

The publisher is not responsible for websites (or their content)
that are not owned by the publisher.

Typeset in Sabon by MacGuru Ltd

Library of Congress Control Number: 2022945406

ISBNs: 9781541600850 (hardcover), 9781541600867 (ebook)

LSC-C

Printing 1, 2023

My senses sing their song of you. The sight and sound
of you are its lyrics, the melody your scent and touch.
Your harmony enfolds me; you are my world.

*– Punkin*

# Contents

# Introduction

*Light, shadows and colours do not*
*exist in the world around us.*

*– Presentation Speech by Professor C. G. Bernhard, Member*
*of the Nobel Committee for Physiology or Medicine*

It's a glorious spring morning in Sydney and I'm full of nervous anticipation as I cross the university campus, heading toward the lecture theatre, where I'm going to be talking to the latest group of students about the senses. I love to watch their faces when I describe the wonders of sensory biology. It's an amazing topic and I want to do it justice; I'm not just relaying information, I'm giving a performance in the hope that my enthusiasm might kindle theirs.

On my way, I cut through a Sydney landmark known as the Quadrangle, the centrepiece of the university campus. The architects added a finishing touch, a subtropical tree, in one corner; each year, as the Southern Hemisphere spring takes hold, this venerable jacaranda tree erupts into bloom, its fragrant lilac flowers calling time on the academic year. Jacarandas across Sydney join in, transforming the city. For a month, the parks and pavements are blanketed with petals. For me, it's the sensory highlight of the year.

As I admire the grand old tree, I can't help pondering how incredible it is that photons of light and molecules of smell weave such majesty. How does my brain access this basic information and transform it in the greatest of all synergies to a perceptual experience?

Though my attention is captured by the jacaranda, I'm aware of a host of other sensations. An Australian magpie is calling from a perch atop one of the buildings that surround the Quad. Its burbling, oddly metallic call sounds like a steampunk version of the songbirds that I grew up with in England. At the same time, I can feel the morning breeze coming in from the Pacific Ocean through the archway on the east side of the Quad. My mouth is filled with the warming flavour of one of the aniseed lozenges that I rely on for a clear voice in each lecture. At the same time, a combination of other senses keeps me upright, while updating my brain on my bodily needs, and keeping me alert to my surroundings.

And this is just a fleeting moment of sensation. The changing stream of sensations provides our perceptual link to the world, a multiplicity of incoming messages that come together to write the autobiography of every second of our lives. For all that our perception seems like a coherent, singular sensory experience, it's a harmony of many distinct, yet compounded, senses. The question of just how many senses still lacks a definitive answer twenty-three centuries since the first reasoned attempt was made.

*

The Greek philosopher, Aristotle, is justly regarded as one of the most influential thinkers in history. Sometimes his ideas didn't quite hit the mark, for instance his assertion that bison discourage chasing dogs by firing caustic turds at them, or his intriguing idea that bees are deaf on the basis that he could see no ears. Notwithstanding the occasional misstep, his legacy is extraordinary. It's been said that the science of biology sprang from his labours and many things that he described over 2,000 years ago have stood the test of time. Indeed, it was Aristotle who is credited with the discovery, if that's the right word, that we have five senses (or, more formally, sensory modalities): vision, hearing, taste, smell and touch. Often, Aristotle gets a bad rap for this, largely because it seems to state the bleeding obvious. In his defence, this was only a small part of his insightful and ground-breaking theories about perception and how the senses combine to provide us with our experience of the world. Nonetheless, poor old Aristotle's name is typically drawn into the fray any time the question is asked: 'how many senses do we have?'

This rule of five is still the basis for our early education in the senses, yet it's some way from the whole story. We certainly have more than five and depending on how we slice and dice the different categories, we might have as many as fifty-three. Touch, for instance, is a composite of multiple different senses that could be subdivided, then there are others such as equilibrioception (the sense of balance) and proprioception (our sense of our body's position) that lie outside the original five. Putting a precise number to the senses, though popular as a quirky topic of debate, isn't especially helpful. Nevertheless, it's important to know what we mean when we describe something as a sense.

Generally speaking, a sense can be defined as a faculty that detects a specific stimulus by means of a receptor dedicated to that stimulus. For example, when light enters our eye, it is absorbed by a molecule known as a retinal, which is found within the light receptor cells of the retina. The light's energy causes the retinal to perform a tiny molecular contortion, in turn setting off a chemical chain reaction that ultimately produces a minute quiver of electricity. It's this tiny zap that gets transmitted along the optic nerve to the waiting brain, which interprets the message and countless others that arrive simultaneously from neighbouring receptors to provide us with the visual sensation of the light. This process of converting a stimulus into a signal that the brain can understand is known as transduction.

Taste receptors, meanwhile, coat our tongues, the inside of

our cheeks and the very top of the oesophagus. Give them a molecule and milliseconds later, they'll be telling the brain all about it. We also have taste receptors sprinkled around the body in places such as the liver, the brain and even the testes. This latter revelation, from a paper published in 2013, gave rise to a fad among young men to dangle their balls in such things as soy sauce, with some even claiming to have registered a savoury hit. The thing is, though taste receptors may be found in such extraordinary places, they're not organised into taste buds and nor are they wired to the brain in quite the same way as the receptors in our mouths, so they don't deliver the experience of flavour. The net result is that the participants exposed themselves not only to condiment-covered gonads but to accusations of wishful thinking. Notwithstanding the bowls of ruined soy sauce, a sense is only a sense if it involves not only specialised receptors, but a functioning information highway to the brain's sensory cortex. Yet though the nervous pathways of our senses lead inexorably from receptors to brain, it would be wrong to conclude that the brain is merely a computer, neutrally receiving and decoding input.

*

The brain is the seat of all your knowledge, emotions and personality; it's the home of your innermost thoughts and the place where you experience everything in your life. Situated safely within the protective capsule of the skull, the brain sits in a carefully controlled physiological equilibrium. It has no sensations of its own, yet this is where all our experiences occur. Supplied by a vast and complex network of connections to the sensory organs, the brain receives the equivalent of terabytes of information every second. It processes and interprets all of this information almost instantaneously, meshing together input from different sources in a seamless computational feat that has no equal. The result of all the work that the brain does in sifting, ordering and processing the incoming information is known as perception. But this is a far from passive process. The brain doesn't simply collect and organise data, it actively regulates and conditions. Signals from the outside world are interpreted and layered with biases, prior expectations and emotions. This integration of sensations and sensibilities plays a powerful role in our perceptions.

Many years ago, on the only occasion that they set foot outside of Britain, my grandparents travelled to Vienna. It had always been my gran's dream to visit there, to revel in the beautiful city, to see its architecture, to taste Sachertorte, to hear the famous waltzes in their birthplace. Later she recounted how they'd rounded a building and come across the famous river that bisects the city.

'Look, Jim! The Danube!' she called out in her excitement. 'They say that if you're in love, it looks blue!'

My grandad wasn't a man easily stirred by poetry. His York-shire vowels as flat as the cap that he habitually wore, he replied tersely, 'Looks bloody brown to me.'

While common sense might dictate that the waters of such a major, industrialised river would not ever resemble an azure, sylvan pool even to the most hopeless romantic, there is a nugget of truth to this. When we're emotionally aroused, activity increases in the brain's visual cortex and what we see becomes richer and more brilliant, even if not necessarily bluer. As for my grandad, his sensations on that trip were likely guided by his attitudes. Our mindset, to some extent, influences neural activity in our brains so that we see what we expect to see.

Ultimately, the convincing perception of reality that we each enjoy is actually a complex but brilliant illusion. This, more than anything else in discussions of the senses, causes people to baulk. We think of ourselves as rational, discriminating creatures, so how can our immediate experiences be illusory? To illustrate this, we can use a simple example. I have a mug of tea in front of me as I write. If I were to ask someone to inspect it closely and describe it, they might tell me the colour of the mug and its contents, that it smells of tea, that it's hot. If they took a sip, they might tell me that it tastes slightly bitter, milky, and overall, well, like tea.

Their experience of my mug of tea would seem entirely and objectively real to them, and they'd take their reality to be identical to mine. Yet while our sensory experiences of the tea would overlap extensively, the overlap wouldn't be complete. Our appreciation of the subtleties of colour might differ. Likewise, the smell and taste of the tea would be different for each of us. If the other person had recently come in from the cold, the tea would feel warmer to them.

In addition, our feelings colour our perceptions. Perhaps the other person is from the Middle East and is appalled at the idea of putting milk in tea. If this were so, their response to the mug of tea would be shaped, in part, by their cultural judgement. The experience feels real to each of us, yet no experience is objectively correct. That doesn't stop people trying to argue that their subjective perception trumps that of others.

This shading of different realities is only the start of the great illusion. It gets more fascinating, and much weirder. It's one thing for people to allow that there might be an alternative perspective on colour, for example, but it's quite another for people to accept that colour doesn't actually exist outside of our brains. Not only is there no colour, but there's also no sound, or taste, or smell. What we perceive as red, for example, is just radiating energy with a wavelength of around 650 nanometres. There's nothing intrinsically red about it; the redness is in our heads. What we think of as sound is just pressure waves, while taste and smell are no more than different conformations of molecules. Although our sense organs do a splendid job of detecting each of these, it's the brain that construes them, converting them into a framework for us to understand that world. Valuable though this framework is, it's an interpretation of reality and, like all interpretations, it's subjective.

The seamless conjoining of all our sensory information into a single, integrated experience is no mean feat and to achieve it, the brain relies on a little trickery. For instance, it has to compensate for discrepancies in the time it takes to process the different senses: vision, being so data rich, takes a fraction longer than the other senses. That's why even in the twenty-first century we begin sprint finals with a starting pistol, rather than traffic lights. The fact that a pistol is used isn't about tradition, some anachronistic nod to our frock-coated forebears, it's simply because athletes, like the rest of us, are slightly slower to react to the light than the sound. Our sensory synchronisation is only possible because the brain imposes slight lags on different senses to make everything line up. Moreover, everything we experience has already happened by the time we register it. To keep up with the real world and to compensate for this slight delay, the brain has to predict movements. If it didn't, we'd be hopelessly out of sync and clumsy.

With so much information flowing in, demanding immediate attention, how does the brain manage to keep up with it all? The answer is that it doesn't. It filters and winnows the information in its perpetual quest for what's important. It's especially attentive to novelty and change; most of the sensory information that we constantly gather never make it past the periphery of our attention into our consciousness. If you're sat down now, you're not likely to have registered the pressure of the chair against your back, or the clothes against your skin – at least until you read this sentence. This isn't the brain being lazy, but rather it's just separating the important from the irrelevant. The downside is that the brain often misses subtleties, which is how dextrous magicians manage to fool us so consistently.

This illustrates the bottleneck between sensation and perception, between collecting information and processing that information to the point that we become consciously aware of it. This is particularly important in vision. The brain seeks patterns and cuts corners by using a template, known as the internal model, of what it can expect to sense based on its experience of what it has sensed before. This can be incredibly useful in that it allows the brain to work with incomplete data, conjuring a full picture from fragments.

However, it's also the reason that we're subject to illusions, and vision in particular is subject to being fooled. Take for instance the well-known film of the rotating theatrical mask. As we watch the mask slowly revolving, we might first see the convex surface of the mask and that's easy; faces are the human brain's bread and butter and everything makes sense. But what happens when we see the mask's concave side? The brain turns it inside out, so we invariably see it as convex surface, like every face we ever see. Even though we know that what we're seeing is hollow, the brain's internal model overrides our reason.

*

The dominant role played by the brain in perception means that we might envisage it as the conductor of our sensory orchestra, coordinating and integrating the separate inputs into a coherent, rich single experience. But without an orchestra, there's no point to a conductor. The brain exists only because there is sensory information to process. In answer to the age-old question of 'which came first?', the senses are the egg to the brain's chicken. Indeed, plenty of organisms get by without a brain, yet many of those can still perform basic sensing. Imagine a bacterium, far smaller than can be seen with the naked eye, seeking nutrients amid the expanse of a cup of water. Its hair-like whip of a tail spins, describing microscopic circles that push it along like a boat's propeller. The bacterium has no goal in mind, but it can detect chemicals in the water and follow them to their source. It locates a faint trace of sugar, a welcome meal for a hungry traveller, and moves toward it. As it approaches, however, it senses a new chemical, a protein, that indicates trouble in the form of another organism. Reflexively, the tail spins again, this time in the opposite direction and the bacterium changes course. This story of how bacteria such as *Escherichia coli* track nutrient gradients is simple, yet it describes the operation of something fundamental: the very first sense to emerge.

Life evolved in water some 4 billion years ago. The first organisms were static, unable to move except with the assistance of currents. Staying exactly where you are isn't the most satisfactory arrangement, however. The ability to seek out pastures new allows an adventurous microbe the chance to exploit new and untapped resources. Cyanobacteria, among the first life forms to appear, achieve their ambitions of mobility in various ways. Some squirt out little jets of slime to propel themselves. Bacteria glide, crawl and swim as a means to relocate. Mini migrations such as these are much more effective if organisms are able to navigate. Chemical gradients are one property of the physical world that provides them with their bearings. Light is another. Photosensitive proteins, such as rhodopsin, absorb light and as they do so, they undergo a chemical reconfiguration that is the basis for detecting the sun's rays and the sustaining energy that they provide.

These foundational steps in the evolution of complex, sensory life were accompanied by another – the ability to detect changes in pressure, otherwise known as mechanosensitivity. Bacteria have channels in their outer membranes that open in response to pressure. Essentially, these are what stops them bursting after overdoing it on the pudding, they're what allow the bacteria to match the pressure of their inner selves to the outer world. It's been speculated that these sensitive channels were the forerunners of our own, more elaborate mechanosensation. Certainly, by the time we get to more sophisticated organisms, for instance Protists, like *Paramecium*, we can see that they respond to touch. Like bacteria, *Paramecium*'s entire body amounts to just one living cell, but giving it a gentle tap causes its internal pressure to change and it responds by zipping away in the opposite direction. Incredibly enough, this simple riposte to mechanical stimulation is what eventually gave rise to hearing and touch, just as light detection was the starting point for vision, and bacteria's ability to track chemicals ultimately yielded our senses of smell and taste. These advances occurred billions of years ago in the simplest creatures and it's a sensory legacy that has been passed down to each and every branch of the tree of life.

Across evolutionary history, organisms have climbed a sensory ladder, with each new rung offering an extraordinary advantage to those that ascended it. The crucial currency for these advances is information: about the environment, about predators and prey, about competitors and potential mates. Our senses were bequeathed to us by ancient organisms following gradients in a primeval swamp, and ultimately these senses were the driving force behind the evolution of the brain.

In fact, the normal workings of the human brain depend on sensory inputs and in the absence of these, strange things start to happen. Recently, I visited a sensory-deprivation chamber in Sydney's eastern suburbs. For the most authentic experience, I was told, I'd need to be fully undressed, to avoid the sensation of clothing against my skin, which might put a barrier between myself and the bliss that awaited. And so it was that I found myself stark naked and self-consciously stepping into an egg-shaped pod, before pulling the lid closed and embracing sensory oblivion. I lay down, my weight supported by a shallow pool of super-saline water at the same temperature as my blood and with ear plugs to still the faint noises from without.

At first, my main emotion was a kind of fretting boredom, my mind chiding me like a fractious child for the withdrawal of stimulus. Once that passed, it switched to stand-by and I relaxed, but in the absence of anything to see, my mind started to conjure things – flashes of light, geometric patterns that fizzed to life and then shrank to nothingness. This is formally known as the Ganzfeld effect, or more evocatively, 'the prisoner's cinema'; it's been experienced by miners trapped in the dark underground, and by polar explorers whose entire visual field may consist of a uniform white. In Ancient Greece, there are records of philosophers descending into caves to induce these hallucinations, in the hope of gaining insight. Given time, the light show can sometimes develop into more fantastical waking dreams. Underlying all of this is the brain's frantic efforts to build its internal model, even though the sensory information it needs to construct that model has been cut off. The results are odd, though to some they can feel disturbingly real. In normal life, for most people, this internal model provides the brain's sensory framework, an illusion that it augments and updates as information comes in. It's this fantasy that paradoxically provides our experience of what we call reality.

*

But what is reality, and more generally, what does it mean to be alive? However we might try to answer this, it's fair to say that even our most eloquent attempts fall short of fully conveying the ridiculous, magnificent, miraculous experience of being. Our senses are at the heart of all this wonder. They are the interface between our inner selves and the outside world. They equip us to perceive beauty, from great art to the grandeur of the natural world, and to appreciate a sip of an ice-cold drink, the sound of laughter, the touch of a lover. Senses are, in short, what make life worth living. Our sensory receptors harvest a multitude of textures, pressure waves, patterns of light and concentrations of molecules to feed myriad pulses of electrical information, like an army of hyperactive stenographers, to the brain, which decodes, organises and, ultimately, weaves meaning. This extraction of meaning from the jumble and chaos of physics is what makes us, us.

My own understanding of the senses is forged from the perspective of a biologist and through my studies on the sensory ecology of a variety of different animals at the University of Sydney, and before that at universities in the UK and Canada. My research has examined which stimuli guide the behaviour of creatures, from insects to whales, and thus how each experiences its own domain. The greatest challenge that comes with this is to try and set aside my human-centric biases, to comprehend things from very different perspectives. While I can never perceive things quite as other species do, I can at least attempt to shed the certainties of my sensory experience and endeavour, so far as I can, to see the world through their eyes. It's this process more than anything that ignited my passion to know more about the senses, not only in other animals, but in us.

As a biologist, it's essential to understand why it is that evolution has equipped us with the senses that it has. To do this, I delve into the sensory lives of creatures – from the mammals with whom we share a recent common lineage to those that are far distant from us, such as crustaceans and even bacteria – to learn about the origins of our senses and the ways in which our experience differs from theirs. While this is primarily a book about the human senses, by exploring the sensory worlds of other animals we can gain a deeper appreciation of our own.

In my quest to understand the senses in the broadest possible way, however, I soon realised that I had to reach beyond my own field. The senses are not just about anatomy and physiology, for all that some of the drier textbooks may present them as such. An approach that restricts itself to processes doesn't begin to convey the wonder, or the deeper meaning, of the senses. Freeing myself from the constraints of a purely biological viewpoint, I immersed myself in research from disciplines as diverse as psychology, ecology, medicine, economics and even engineering and I delved into the question of how thoughts, emotions and culture shape, and are shaped by, our sensory world.

My challenge was not only to understand the senses but to place them in the context of our lives and it's the challenge that has inspired me to write this book. While I don't neglect the underlying biology, my goal is to examine our senses in the round. For this reason, I leave the more detailed biochemistry, molecular and cell biology to other, more specialist books. Instead, I examine not only how we sense, but why. I'll delve into the fascinating questions of how we each differ in our sensory experiences and where these differences emerge from. I explore how our senses have shaped humankind and I look to the future, to predict how the senses will influence what is to come.

I've arranged the book by devoting a chapter to each of our five primary senses before turning my attention in Chapter 6 to a host of underappreciated but crucial senses. But while such an approach has the benefit of neatness, it runs the risk of implying that each sense is separate and segregated from the others. This, as I show, is far from the case; all of our senses are interdependent and fascinatingly so. As a result, I examine the many interactions between the different senses throughout the book, and especially in the final chapter, where I explore how our brains weave the miraculous tapestry of perception from a medley of sensory inputs.

When I began this book, I came to it full of enthusiasm for all aspects of our sensational existence. The research that I've done in the intervening years has only amplified my appreciation of this incredible topic. The Nobel laureate, Karl von Frisch, once described the process of learning about a subject for which you nurture a great passion as being like a magic well: the more you draw from it, the more it fills with water. I wish you, the reader, a similarly wonderful experience as you dive into the extraordinary world of our senses.

# What the Eye Sees

*A view comprehends many things juxtaposed,
as co-existent parts of one field of vision. It
does so in an instant: as in a flash, one glance,
an opening of the eyes, discloses a world of
co-present qualities spread out in space, ranged
in depth, continuing into indefinite distance.*

– 'The Nobility of Sight', H. Jonas

Sight is sometimes regarded as the ultimate arbiter of truth. When we're told of some fantastic episode, we might reply that we need to see it for ourselves. Yet what we see isn't reality, it's a narrative created by the brain. Subconsciously, the brain takes the raw input from our eyes and it freights the raw input with meaning, filtering the observations and subjectively ascribing qualities and biases, filling in gaps as it goes. Most of the time we're unaware of this, investing our recollection of what we've seen with confidence and certainty, as in the phrase 'I saw it with my own eyes!'. This reliance on vision represents a degree of overconfidence since sight is the sense most prone to being tricked. We even try to fool it for ourselves, for instance when we wear 'slimming' colours, or when interior designers resort to tricks of the *trompe l'oeil* variety.

The sham only begins to reveal itself when we experience an illusion. One of the most basic forms of this is known as the Müller-Lyer illusion, where two identical lines can be seen, usually presented parallel to one another. Both are bookended by a pair of V-shapes; one has the Vs arranged so it looks like a double-ended arrow, while the other has them pointing inward. The fact that the Vs stick out further from the latter makes the line seem longer, even when we know that isn't the case. It's a simple example, which exploits the fact that how we see things depends on the visual context in which we see them.

Context isn't everything, however. The autokinetic effect describes how points of light might seem to move as we look at them. The German scientist and philosopher, Alexander von Humboldt, wrote about 'swinging stars' he claimed to see moving in the night sky. You might experience the same illusion of movement if you gaze at a star, particularly on an evening when relatively few others are visible. It's perhaps understandable how some people interpret such apparent celestial agitation as proof positive of visiting alien ships. But the most compelling example of the effect comes from studies where participants are asked to look at a single, stationary point of light on a screen and are told that the light is moving in a certain direction. Primed with this information, the participants most often agree that the light is moving as suggested. Best of all, in another, similar study, experimental subjects were informed that the light would spell out a specific word, although they weren't told what the word was. Of course, the static light couldn't spell out any words; if the participants saw anything, it could come only from their imaginations. Yet when asked, many insisted they had seen a word, and in some cases refused to disclose what the word was because it was rude.

The brain gathers only the gist of what's in our visual field beyond the thing that we're concentrating on. That's why we suffer from phenomena such as inattention blindness, most famously evidenced in a video that went viral on social media some years back. Asked to count the number of passes made between a group of basketball players most people were so caught up in the task that they didn't spot a person in a gorilla suit walking through the frame. We're attuned to the big picture and we're good at recalling the essence of what we see, but very few people have the capacity to recount the details of a given scene, something which makes eyewitness testimony a rather hit and miss affair. We look, but we don't always see. Even with all of these flaws and inconsistencies, we are arguably, more than anything else, a visual species. Oddly, however, it's a sense that we very much have to grow into.

＊

It is hard to think of a more profound and intense experience than that of being face to face with a loved one, silently contemplating one another. We overwhelmingly look at, or into, the eyes and we're conscious of seeing and being seen. When fathers gaze at their newborns, evolutionary theory tells us that they're looking for something: a resemblance. Across different cultures, nonpaternity rates average a little over 3 per cent. In other words, around one in thirty fathers of newborns aren't the father at all. Perhaps it's for this reason that mothers (who, after all, can be reasonably confident of being the parent) are four times as likely to point out their baby's resemblance to the father than the other way around. A study carried out in 2009 asked people to rate the similarity between fathers and their children and then followed up by asking the mother to give feedback on how good a dad their partner was. The results were remarkable. The greater the resemblance, the more effort the fathers put into raising their children. Overall, the more confident a man is that the child is his, the more he tends to invest in that child, and one of the most important elements in generating that confidence is the similarity that the father perceives in his child's appearance. It has to be said that as I regarded my son on his first day at home, I wasn't aware of making any such assessment. All I knew was that this dribbly, incontinent bundle of gurgles was the most wonderful thing I'd ever seen.

It's perhaps fortunate for my son that at the age of less than a week and with my unlovely face looming over him, his vision wasn't great. Much of what he'd have been able to see would have been a blur. This is common to all newborns; the clarity and definition of whose vision is only around 5 per cent of that of a normally sighted adult. They can see faces, but only at a range of around 30 cm, which handily is about the distance from their mother's breast to her countenance. Faces are arguably the most important things that we, as intensely social animals, have to recognise. The basics of this ability are present even before we're born, particularly the tendency to relate to a rough configuration of two eyes abutting a nose with a mouth below. Third trimester foetuses respond to light patterns shone on their mum's belly and when an arrangement of dots and lines is used in an approximation of a face, this holds their attention for much longer than other, similar constellations.

Our tendency to tune into this most basic of facial contours – essentially two dots and a line – is why we're so prone to seeing faces in clouds or on the fronts of cars. Fortunately for us, our ability to recognise faces is a little more complex than this, but the way that we achieve this sheds some light on why two dots and a line can at least begin to fool us. Recognition is achieved by a network of neurons in the brain. Each group of cells within this network attends to a specific characteristic of a face, and then in collaboration, the groups build a composite picture that we use for identification. Among all of the complexities of a human face, however, it's the crucial pattern of eyes, nose and mouth that anchors our perception and provides a kind of mental canvas onto which we can map the other features.

We, like many other mammals, are born incomplete as sensory animals. Our genes provide a kind of rough draft of the neural equipment needed for perception in our brains. This rough draft is shaped and honed by experience, especially in the critical first weeks and months of our lives. Missing out on this experience can lead to lifelong deficits. Mice reared in the dark never fully establish sight to the extent that those with a more typical rearing environment can. The same is true, sadly, for people who lose their sight as infants and later have it restored through surgery. In visual terms we're born as beta versions, stimulating and reorganising the brain simply by looking around us. It takes around six months to fully hone and train our sight, which is testament to the staggering intricacies of human vision. It was not always this way. Far back in evolutionary history, what we now think of as sight began as merely the ability to register light.

*

Of course, we can never know exactly how light detection evolved or what form it took in ancient times, but it seems likely that it wasn't too different to the equipment that some modern single-celled organisms now possess. Photosynthetic bacteria gain their energy from the sun – thanks to light-sensitive proteins they can, at some level, register the presence of light, but the problem for many of them is that they don't know where it's coming from. Consequently, they blunder around their environment until they chance upon their equivalent of a sunny glade. The alga *Euglena* boasts a much greater degree of sophistication. It's just one cell, but it can sense light and it has a whippy little tail known as a flagellum that it uses to propel itself towards it.

Guided by similar light-sensitive pigments, a sapling detects a gap in the forest canopy and hurries up to meet it. If the light strikes the sapling at an angle, part of the plant will be in shade. The cells on the shady side respond to this solar snub by extending and elongating, which has the effect of bending the tip of the plant directly towards the sun. Some fungi, such as *Pilobolus*, take this a step further. *Pilobolus* specialises at growing in the rich, moist environs of animal turds, and like all good parents, they look out for the interests of their offspring. For the next generation of *Pilobolus* to prosper, they have to be eaten as spores by an herbivorous animal and subsequently evacuated along with some ready-made fertiliser. The problem, however, is that grazers tend to avoid feeding too near dung. What the adult fungus must do is find some way of flinging its spores into a different neighbourhood and this is where it's essential for them to be able to detect the sun.

Like our sapling, these so-called hat-throwing funguses can sense light and turn towards it. To assist them, they have an important refinement. At the very top of their slender stem, they have a transparent sac of water. This enclosed globule of liquid acts like a lens, focussing the sunshine onto light-sensitive cells below and allowing the fungus to register the sun's rays more effectively. In the early morning, when the sun's light hits the fungus from the horizon, the fungus bends towards it and does the job it's named for: it throws its 'hat' – in reality, a parcel of spores sat right on top of the fungus's makeshift lens. The water in this lens is at such high pressure that when the sac ruptures, the package of spores is subjected to a G-force twice that of a bullet fired from a rifle. By aiming at the rising sun as it's low in the sky, the fungus ensures that the kids go sideways rather than straight up. So the spores are propelled to a bright new future, well away from the parental pile.

Surprisingly enough, this simple approach to light detection also occurs among animals, many of whom can detect changes in light with their skin. When a shadow passes over a sea urchin, the prickly (but eyeless) little creature realises it might just be about to come under attack and bristles its spines in response, declaring itself up for the fight. Shine a light on the tail of a lamprey, or on the larva of a fruit fly and they scoot off to find shelter – in both cases, a response that's independent of eyes. Pigeon chicks sit up and beg for food when the light above them changes, a cue that they're hard-wired to associate with the arrival of their parent. Amazingly, they do this even when they're wearing a hood that completely covers their head, but they don't do it when they're attired in a cape that blocks the light from their entire body. The responses of all these creatures are achieved with the help of photosensitive proteins in the skin that register the presence of light.

Relatives of these proteins can be found in our own bodies. As we awake each morning and open our eyes, light streams in and sets in motion a cascade of events that ushers away sleepiness and gets us ready to face the day. It does this courtesy of an extraordinary protein, known as melanopsin, which can be found sprinkled around various locations inside our heads and within our eyes. When light hits melanopsin, the protein does a little molecular dance that results in the sending of a message to the suprachiasmatic nucleus deep in the brain. In response, the bundle of nerve cells therein shut off production of melatonin, the hormone responsible for preparing us for sleep, and kick-start our bodies into action. Melanopsin is specifically excited by blue light, which is a feature of the backlit screens that we like to gawp at, and it's one of the reasons why it's a terrible idea to take your phone to bed. Reading such a device activates melanopsin, in turn persuading the brain to keep you awake.

Impressive though melanopsin is, it doesn't enable you to see. Its job is simply to register the presence of light; it's a long way from detecting light to vision, and for this you need eyes. Gliding around on a carpet of ooze at the bottom of a pond, tiny creatures known as flatworms carry the most rudimentary versions of these organs. Towards the front of the flatworm's body is a pair of eyespots, clusters of light-sensitive cells situated within little cup-like depressions. Like many shady characters, the flatworm likes to stay out of the limelight. Armed with its eyespots, and particularly the directional shading provided by the cup, it can tell where light is coming from and can use this to remain in the shadows.

It's perhaps forgivable from a human perspective to feel a little underwhelmed by the mere ability to detect where light's coming from. After all, the flatworm's visual abilities don't seem to represent much of an upgrade on what a fungus can manage. But pause for a moment. Just being able to orient yourself according to a light gradient represented a revolution in the history of life of Earth. For a microorganism like *Euglena*, it means the ability to hog the sunniest places and to merrily photosynthesise while less sophisticated competitors lag in its wake. For a flatworm, it means the ability to find shade. Such creatures have a competitive advantage over those that lack their light-finding nous. Ancient organisms that were equipped in this way were rewarded through natural selection; they have more offspring, which tend to inherit the very qualities that made their parent a winner.

Even so, we're still some way away from eyes. More specifically, the bit we're missing is the precise reason that you can read this page: the ability to form an image. How did we get from mere light-sensitive patches to the glorious sense of vision that we enjoy? Our eyes, and the whole visual processing systems of our brains, are so intricate and comprise so many essential components that it seems incredible that they could arise incrementally. In reality, we can hypothesise how this happened and, crucially, as we retrace the steps of more than 500 million years of ocular evolution, we can see examples of eyes in various stages of complexity in the animals around us.

Starting with the flatworm, the deeper the pit that contains their eyespots, the better the job this pit does of casting a shadow on the light-sensitive cells that reside within. If the opening of that pit is relatively narrow, what you effectively have is something like a pinhole camera. No lens is needed, but the effect is that light entering a narrow aperture will project a simple image on the opposite side. Admittedly, the image isn't fabulously clear, but it is nonetheless an image and animals such as abalone and nautilus rely on this arrangement even now. Building on these foundations, the development of the eye gathers pace. It develops a transparent covering of skin, perhaps originally to keep pathogens out, that evolved in time to become the cornea. The lens, too, developed from skin cells with high concentrations of transparent proteins known as crystallins. The sensitive cells that perform rudimentary light detection in animals like flatworms developed into the exquisite structural array that we know today as the retina. A combination of the lens and the cornea bends light into shape and focusses it onto the retina, allowing us to form a beautifully clear image.

It would be wrong to imply that the evolution of eyes represents some kind of predictable progression of biological achievement from primordial light-detecting bacteria to some pinnacle in modern humans. Many eminent biologists have hazarded guesses at the number of times that vision has independently evolved over the last 500 million years, ranging from a handful to hundreds. Regardless of what the correct number is, it's certainly true that there's a dizzying diversity of eyes in the animal kingdom, some of which are demonstrably superior to ours. And like so many products of evolution, our eye is assembled from a hotchpotch of available components and represents a series of compromises. The genes involved in coding the development of the eye are scattered around our genome rather than being collected together at a single point. What's more, those 'eye' genes have histories that extend way back to a time before eyes existed. The original role of some of these genes was in coding a kind of cellular stress response, comparable to that which causes our bodies to tan following exposure to ultraviolet light. The bottom line is that the genetic equipment needed to build an eye didn't appear out of nowhere, but rather involved taking a bit from over here, another from over there and so on.

The result is an excellent eye, for sure, but it's by no means flawless. The most obvious glitch is our strange back-to-front retina. The blood vessels and nerves that respectively supply the retina and connect it to the brain are on the side that faces toward the outside world. As a result, we have a blind spot where the optic nerve wires straight through the retina. And if blood vessels become clogged, or leak, this can interfere with the passage of light to the retina with the effect that vision is blurred or blocked. Biological engineering often bears these hallmarks of imperfection.

*

Imperfect though the eye may be, it's an entrancing organ to behold. If you look closely into someone's eyes, the chances are that you'll notice a beguiling pattern of colours around the iris, but also a tiny reflection of yourself in the deep black of their pupil. It's this that gave the pupils their name, from the Latin *pupilla*, or 'little doll'. Perhaps more than any other part of the body, the pupils are a giveaway to our mood. They dilate when we're aroused and as such, they act to communicate our interest. It's why some poker players elect to wear dark glasses as they play. The response is involuntary – there's not a great deal we can do to mask this and for that reason the pupils are to some extent an honest signal of how we're feeling. Though it's something we're only subconsciously aware of, when we interact with people whose pupils are enlarged, we regard them as warm and friendly, because they appear to be fascinated by us. In past times, women used to exploit this quirk of human nature by dousing their eyes with tincture of deadly nightshade. The effects of doing this are twofold. First, the nightshade blocks the muscles that contract the pupil, which has the effect of making the pupil alluringly large. And second, it blurs vision, and makes it difficult to focus. In consequence, women who adopted this technique would appear to be powerfully attractive right up until they stood up, tripped over the cat and went face first into the chaise longue. Still, the upsides to this rather hard-core treatment were such that the use of it by the ladies of Renaissance Italy was commonplace and their enhanced appearance gave nightshade its alternative name, belladonna, or 'beautiful woman'.

Toxic shortcuts aside, the way your pupils respond when you look at others really does give a tell-tale indication of sexual orientation, although it differs according to your gender. A study that examined the pupils of subjects while they watched raunchy film clips reported that the way they responded correlated with what those subjects described being turned on by. Heterosexual men's pupils dilated more in response to seeing a video of a woman rather than one of a man, while the pattern in gay men was reversed. For women, the picture was more complex. While gay women's pupils reacted more strongly to other women, straight women's pupils responded more or less equally to either sex. This latter pattern has attracted some interesting explanations. Based on conversations I've had with female colleagues, I think it might reflect the greater nuance in female sexuality and female perspectives generally. Rather than suggesting that heterosexual women are secretly bisexual, it may be that while they find the male actor physically attractive, they also empathise and identify with the woman in the video, independently of any sense of being attracted to her. Other explanations are available ...

Just like a camera, the eye has to do more than simply let light in, it has to bend it in such a way that it becomes focussed. This job is done with a little teamwork between the cornea and the lens. The cornea sits on the surface of the eye, above the iris and pupil, protecting the delicate structure from damage, while the lens is positioned on the inside of the eye, behind the iris. We're used to thinking of the lens as being the senior partner in this, but around two thirds of the eye's focussing work is done by the cornea, before light even reaches the lens. Nevertheless, it's the lens that's in charge of the fine-tuning.

Two hundred years ago, the scientist Thomas Young was puzzling over the question of how the eye brings objects into focus. One idea was that the eye itself changed shape, in particular getting longer or shorter from front to back to change the distance between the lens and the retina, just as cameras do. But how would you test this? Young did what I'm sure any of us would have done in his position: he stuck metal keys in his eye socket to clamp his eyeball. His thinking was that if he could prevent his eye from changing shape, he'd learn whether it was this that allowed him to focus. But with his eye held captive in this bespoke torture apparatus, he looked around and realised everything was in focus, gaining both greater understanding and a sore eye. Having ruled out shape changes in the eyeball and in the cornea, Young arrived at the correct conclusion, which was that it is the lens that changes shape to allow us to focus.

Despite his impressive commitment to the experiment, Young was frustrated in his attempts to work out just how the lens changes shape. We now know that our flexible lens is pulled into different shapes by surrounding muscles. Contracting these muscles makes the lens transform into an almost round shape, which bends light dramatically and is exactly what's needed when you need to focus on something close up. As you get older, the lens becomes less flexible and the muscles that control it become weaker, making it harder to see nearby things. Until we reach the age of about thirty, we can focus reasonably well on objects that are only 10 cm from our face. This so-called near point then recedes until, by the time we're aged sixty, it's 80 cm away, necessitating either long arms or special glasses. Going the other way, plenty of people of all ages suffer from short-sightedness. Part of what's happening here is that the eyes focus light onto a point in front of the retina rather than directly on it – it's sometimes said that myopia is the result of having elongated eyeballs.

Lining the back of the eye is the endlessly amazing retina, a thin strip of multilayered tissue at the back of the eye that translates incoming light to nervous signals. It's amazing not only because of what it does, but also because of what it is. The retina is neural tissue, so although it's on secondment to the eye, it's still technically your brain. Indeed it's the only part of the brain that can be seen without cutting into the skull. If you removed the retina and flattened it out, it would cover an area of less than a quarter of a credit card yet crammed within that space there may be more than 100 million photosensitive cells, dedicated to the collection and communication of information from light.

It's a little over 100 years since this incredibly complex, multilayered structure was first properly described, by a man whose early life gave no hint to his later greatness. Santiago Ramón y Cajal was born in northern Spain in 1852. His early life was characterised by a rebellious streak, which saw him barred from a series of schools and perennially dodging the local police. To his father's exasperation, Cajal dedicated his energies to fighting with other boys and devising weapons for this purpose. Rather impressively, this culminated in his construction of a homemade cannon, which he used to destroy his neighbour's door and for which the amateur artilleryman spent a few days in jail. He was saved from delinquency by his passion for art, particularly painting and photography, and science, to which he devoted himself outside the strictures of the classrooms that he so detested. You can see the confluence of art and science to wonderful effect in the incredible drawings that he produced of the nervous system, particularly the retina. Most important of all, he discovered that the cells in the retina connected with one another to make an intricate communication network, a kind of biological precursor of the sensors found in a digital camera's scanner, that could collect detailed visual information and relay it to the brain.

The problem for Cajal was that his findings ran contrary to the prevailing scientific viewpoints on the nervous system. This might have represented a problem, were it not for his inexhaustible willpower. Frustrated by the lack of recognition accorded to his work, he travelled to Berlin, then the epicentre of world science, on a mission. Once there, Cajal didn't so much introduce himself to Albert von Kölliker, the most prominent scientist in his field, as drag him to look at his new findings. Whatever the propriety of this, it worked. Von Kölliker became his most enthusiastic supporter, and Cajal's work laid the foundations for our understanding not only of the retina, but of neuroscience more generally.

Cajal's exquisite diagrams of the retina show not only its many layers, but the two critical types of cells that detect light. These cells are named for their basic shape: rods and cones. Each contributes to vision in a different way. Rod cells don't allow us to perceive colour, they give us a sense of light and dark in monochrome – a kind of fifty shades of grey. Nonetheless, they're more sensitive to light than cones, so are particularly useful in low-light situations. By contrast, cone cells are sensitive to specific wavelengths of light and so give us our perception of colour. The way this works is quite ingenious. Humans typically have three different types of cone cell, each of which specialises in a different wavelength of light: short, medium or long, which roughly translates to blue, green and red. All of the colours that we see are produced by mixtures of these three colours. This is the basis of what's known as the trichromatic theory, an idea that was anticipated by our eyeball-squeezing friend, Thomas Young. It's also why each pixel on the screen of your TV, or your iPhone, has three little dots of colour within it, allowing the screen to mix these in various ways to produce the full range of colours. It's not possible to see these in the screen under normal conditions, but put a tiny drop of water on the screen and look again. The magnifying effect of the water droplet lets you see the pixels and their colours.

As schoolchildren, we're told about the primary colours – red, yellow and blue – those that can't be made by mixing other colours. Strictly, red, yellow and blue are the subtractive primary colours.* Sunlight and the lights we tend to have in our homes are a mix of all possible colours, which is to say they're white. When white light hits something, such as the pigments in paint, or the petal of a flower, certain colours are absorbed – subtracting them

---

* It's actually more correct to say that cyan, magenta and yellow are the subtractive primary colours. That's why printer cartridges bear these names. The colour red is made by mixing magenta and yellow, while blue is a blend of cyan and magenta.

from the mix – while others are reflected and this reflected light is what we see. So when you see a ripe tomato, for instance, its redness emerges because the fruit absorbs all the colours except red, which it reflects. And when an object absorbs all the colours, then that object appears black. But this subtractive scheme only applies when light bounces off objects, like tomatoes, on its way to our eyes. When light comes direct to our eyes from its source, however, things change and we need a different set of primary colours, referred to as the additive primary colours. Coloured lights, such as those you see on a screen, are additive. You start with no light – darkness, in other words – and add colours. You get white light by mixing the three additive primary colours: blue, green and red. Working out how other colours are generated in the additive scheme can be perplexing, largely because those art lessons in early life remain prominent in our minds. For instance, we know that to make orange paint, we mix red and yellow together. By contrast, orange light is a mix of two parts red to one part green.

Our own experience of colour emerges from the way that our brains interpret detailed information coming in from our cone cells. While each type of cone cell is a specialist in a particular colour, each is also responsive to adjacent colours on the spectrum. For instance, our green cone cells don't only get stimulated by green light, they'll register shorter wavelengths like blue, and longer ones towards the red side of the spectrum. The crucial thing is that they only get really excited by green, pinging an enthusiastic message to the brain when confronted by a lettuce leaf, for instance. When they detect a colour either side of green on the spectrum, their passion is dialled down. The same is true for our other cones, and since the range of colours to which each reacts tends to overlap to some degree, the brain can triangulate the information coming in from the three cones to calculate the colour. When we're confronted by something yellow, for instance, our red and green cones fizz with enthusiasm, while our blue cones languish like teenagers at a golden wedding celebration. When encountering a turquoise object, the red cones have little to say, but the blue and green cones are piqued. In each case, the brain interprets the input from the different cone types to provide us respectively with the sensation of yellowness, or turquoise-ness.

Right at the heart of the retina lies an area known as the fovea, from the Latin for 'small pit'. Despite its miniscule dimensions – it's less than half a millimetre across – cone cells are crammed into this little depression in our retina. Not only that, but each of the cone cells here has its own direct connection to the optic nerve, the information superhighway to the brain, and the result of all this is that the fovea provides our greatest visual acuity. This acuity, effectively our ability to distinguish detail, is what's measured by your optician. If you have normal vision, sometimes called 20/20 vision, you'd be expected to be able to read a letter that's about 9 mm high at a distance of 6 m.

Unusually, in terms of acuity as well as motion sensitivity, men tend to outperform women. It's a strange aspect of the human senses that sex differences exist, and in all of our other primary senses as well as in some other features of vision, women have the upper hand. Why would acuity and motion sensing be different? Perhaps it's because millions of years of living as hunter-gatherers placed a premium on the ability of men, who are thought to have done the majority of the hunting, to discriminate detail at long distance and to detect movement of prey animals. The truth is, we don't know. Nevertheless, the differences are small and both men and women have excellent acuity compared to many other mammals. For instance, cats' vision hovers around the limit to be declared legally blind for humans. Dogs do slightly better but still have far lower acuity than us. By contrast, birds of prey have incredible acuity, often well over twice as sharp as our vision, and possibly up to four or five times better, allowing them to see rodents and small animals scurrying around on the ground far below as they soar.

But our impressive visual acuity refers mostly to that which our fovea provides. Outside that area our acuity falls away, which is why your peripheral vision is much less sharp. Retinal cells outside the fovea are less densely packed and have to share their connection to the optic nerve with their neighbours, so the information coming from them is less clear. Rods, which primarily support our peripheral vision, are excellent at detecting the changes in our visual field that indicate motion, but don't supply much detail. When we detect something moving in our peripheral vision, we don't get a clear sense of what it is. An early warning is good in the sense that we can leap out of the way, but it's also the reason we might reflexively apologise to a post box for walking into it.

It's a facet of our sensory systems that when we detect something interesting around us, we present the most sensitive receptors towards it. So when we see something out of the corner of our eye, we reflexively turn to fix that object on the fovea. The vanishingly small size of the fovea means that even viewing something from a distance of 2 m, the area that we see with greatest clarity is only about 4 cm across. Imagine you're looking at someone's face when you're talking to them. Your fovea is sufficient only to see their mouth, or one of their eyes, in high definition. Our visual system has a trick up its sleeve to deal with this: every second, the eye makes dozens of infinitesimal movements, scanning multiple regions of the person's face, which your brain knits together to make the view appear seamless.

If you imagine the retina as a dartboard, the fovea represents the bullseye. The further you go towards the margins of the board, the density of cone cells decreases and their place is taken by rods. Whether your eyes are relying primarily on sharp, full-colour, cone-based vision or the less detailed, black and white perspective provided by the rods depends on how much light there is. As night sets in, there's insufficient light to activate the cones and the rods start to take over. As this transition takes place, the peak sensitivity of our vision shifts away from red along the colour spectrum towards blue. An enjoyable way to experience this for yourself is to enjoy an evening drink in a beer garden and watch how the colours change as it gets darker. The reds fade first, well before greens and blues. If you have the patience to remain, you might be rewarded with a view of the stars scattered across the night sky. Since there's little light, we see them thanks to our rods, which makes them appear white. It's strange to realise that most stars aren't white at all, but a panoply of different colours. To see the full technicolour glory of the stars, you need to amplify the amount of light that reaches your eye from them, which means a telescope. Doing this means that your cone cells kick into action and provide an appreciation of the rainbow of colours above your head. Contrary to our familiar colour coding for temperatures, the hottest stars are blue, or bluey-white, because of the high-energy, short wavelengths of radiation they emit. Cooler stars such as Betelgeuse are reddish in colour.

Throughout our lives, we're bathed in energy that radiates from the stars, including our own sun. We can arrange the different forms of radiation along a spectrum – the electromagnetic spectrum – according to the energy it produces. The Earth's atmosphere screens most of this from reaching us, but crucially two types of radiation do make it through. Low-energy radio waves are one type, which is why we use radio telescopes to study distant galaxies. The other type we call light.

*

The electromagnetic spectrum encompasses a vast range of energies and amid these, there's a vanishingly small segment of radiation, something like 0.0035 per cent of the spectrum, that we can see and which we sometimes refer to as the optical window. Our atmosphere is transparent to these wavelengths and, happily for us, they pass straight through. But in the context of the evolution of vision, these wavelengths were confronted by another filter which narrowed the range even further. This filter is water, or more specifically, the sea. Since our ancestors evolved in the ocean, this narrow range of wavelengths defined the early development of biological sensors. Even when life emerged onto land, where a broader range of wavelengths is potentially available for vision, the die was cast. Ultimately, the colour vision that we have is constrained, its extent determined countless millions of years ago by the biochemical mechanisms that evolved in response to the wavelengths available in water.

Even within this narrow range, there's room for disagreement and this formed the basis of a strange argument recently among the students in my lab concerning a bag. One of them asserted that it was violet, the other insisted the bag was turquoise. Neither side gave an inch; to their own subjective selves, each of them was the sole upholder of truth and the last bastion of sanity. Debates like this seem to suggest that our sensation of colour varies from person to person, that what we see is subjective. More broadly, the dispute opens out into the age-old question: When each of us sees a colour, do we see the same thing? You and I might agree on the label that we give to what we see, we might each describe a ripe tomato as red, but do we both see the same colour? The answer to this question is elusive, partly because colour is an illusion, it doesn't really exist. The ripe tomato isn't red, it's just reflecting light with a wavelength of 650 nanometres. Our brains convert this input and create for us the perception of red. I can measure the wavelength, but I can't experience the sensation that's going on in your head when you look at the tomato. We can't know how differently each of us sees the world, though it's likely that each of us sees it in a slightly different way. Even with this seemingly impossible barrier to understanding each other's experiences, the question of how we see colour has attracted intense scrutiny and it gives us a fascinating insight into how each of us perceives the world.

We learn to label different colours in infancy. We're guided in this by our parents, our peers and our teachers – there's a strong cultural influence at work. This is the basis of a school of thought known as linguistic relativism, which argues that our language determines our perceptions. A famous example of this comes from the studies in 1858 of William Gladstone, later the British prime minister. His dissection of *The Odyssey* revealed some peculiar aspects of Homer's writing. In particular, Homer described colours in a way that we find strange. He used the word purple to characterise blood, dark clouds, waves at sea and even the rainbow. He described the sea itself as 'wine-dark' and made no reference to the colours blue, green or orange. Why? Gladstone had a ready answer: the Ancient Greeks were effectively colour blind. A hard-line relativist would take a different perspective, that Homer's culture and the words he used defined what he saw – in a sense, colouring his views. Benjamin Whorf, perhaps the foremost proponent of this school of thought put the case succinctly: 'We dissect nature along lines laid down by our native languages.'

Dissecting is such a human characteristic. To make sense of things, we put them in boxes. We do this even when we're dealing with a continuum, as is the case with colour. What we call visible light exists as wavelengths from around 380–760 nano-metres, which we perceive as the colour spectrum. There are no clear breaks in the colour spectrum; the colours merge gradually into one another as we progress along it. Nonetheless, we haven't let that get in our way. In his seminal studies on optics, Isaac Newton identified seven colours – violet, indigo, blue, green, yellow, orange and red – that seemed to emerge when he split white light into its various constituents using a prism. Why seven? Well, Newton wasn't insistent on this point, but it could reflect at least partly a fixation on the supposedly lucky number seven in Western culture that goes all the way back to Ancient Greece and that gives us things like seven notes in a musical scale, seven days of the week, Seven Wonders of the World and seven deadly sins, among others. While Newton's work represented a landmark in the scientific understanding of light and colours, that didn't stop the German scientist and philosopher Johann Wolfgang von Goethe from challenging his assertions and expanding on the idea that the perception of colour is a subjective one that we each experience differently.

Subjective though it undoubtedly is, by studying the language of colour across different cultures, we can get a sense of our similarities and differences. In 1969, the anthropologists and linguists Paul Kay and Brent Berlin published *Basic Color Terms: Their Universality and Evolution*, a landmark book that offered a challenge to the relativist idea that language shapes perception. They argued that how we see and describe colours is universal, innate and independent of culture. At the root of their argument they pointed out that most languages cut the spectrum up in similar ways and along similar lines. The words are different, but the colours they describe are largely the same, the implication being that all people see colour more or less alike.

Another fascinating insight from this work is how certain patterns are revealed in the development of terms for colour. In particular, most languages have at the very least terms for black and white. When a language has specific names for three colours, the third one is nearly always red. Perhaps this is due to the biological significance of red, both in regard to indicating an injury, and as a means for finding particularly valuable foods, such as berries. Perhaps it's just because red stands out. It is noticeable that red pigments tend to draw the attention of other animals, including fish and birds (though not bulls, ironically). Subsequent colours also tended to be added to languages in a predictable order. After red comes either yellow or green, and they're followed by blue or brown. English has relative newcomers to the linguistic colour fold. Orange for instance, isn't recorded in the language until 1502, which at least partly explains why the robin redbreast is so called, or why we call people with an orangey tint to their hair a redhead. Before it was given this moniker, on the basis of the exciting new fruit that was beginning to pop up in English markets at the time, we resorted to comparative approaches to describe this colour, such as saffron and the rather disappointing 'yellow-red'.

Nowadays, Newton's take on the number of colours in the spectrum is being challenged. Often we replace indigo and violet with purple – as in the Gay Pride flag, for instance – leaving us with a generally agreed six colours, often referred to as the spectral colours. These six are joined by five others: black, white, grey, pink and brown, to make eleven colour terms. That doesn't mean that there aren't others, obviously. Experts argue about the exact number of colours the human eye can perceive and most arrive at a number somewhere between 1 million and 10 million, rather more than the 20,000 words that the average English speaker has in their vocabulary. Despite the obvious limitations, eleven fundamental colours at least gives us a starting point. Besides, their characteristics are reasonably clear in the minds of most people. Many other colours tend to provoke a little more debate. Asked to describe maroon, for example, some will offer reddish-brown, others will veer towards purple. Nonetheless, there are plenty of languages who make do with far fewer specific words for colours than English, and there are others with more.

Before we rush to the conclusion that all people see all colours the same, there are some interesting exceptions that challenge the rule. English speakers are used to thinking of blue and green as different colours, but various other languages have words that can refer to both without making a distinction, including the Japanese word 'ao' and the old Welsh word 'glas'. Speakers of the Berinmo language of Papua New Guinea use a single word, 'nol', to describe both blue and shades of green, so they describe the grass and the sky as the same colour. In addition to this, they have other colours, including 'wor', that includes yellows, yellowy-greens, and some orange. The linguistic boundary between 'nol' and 'wor' occurs at a point that we might simply call green. The upshot is that English has a division – between blue and green

– that Berinmo doesn't, and Berinmo has a distinction – nol/wor – that English lacks.

The differences between the languages in the way that they chop colours into categories provides fertile material for testing. Consequently, Berinmo speakers have seen a steady trickle of linguistics experts come to their part of Papua New Guinea for the last couple of decades. The results of one test in particular catch the imagination. People are shown a colour and asked to remember it. A few seconds later, they are shown two new samples and asked to pick the one that matches the original, memorised colour. So, for instance, the test subject might be asked to memorise a blue colour sample and then be given a blue and a green sample and asked to say which matched the original. The results were unambiguous. Berinmo speakers were much better at matching the alternatives across their nol/wor colour boundary than English speakers, but the latter won the return leg when the colours spanned the green/blue boundary. In a similar way, Koreans are more effective than English speakers at identifying different shades of green. The Korean language has fifteen distinct colour terms, to English's eleven. One of the boundaries that Korean recognises is that between *yeondu* (yellow-green) and *chorok* (green); both would be described as green in English. The results of this and similar studies support the idea that language is instrumental in shaping our perception of colour. This conclusion is given further weight by the amazing findings that we can only effectively categorise colours (or we're much better at categorising them, depending on which study you read) when we see them with our right eye. Because of the way that our optic nerves cross before reaching the brain, it is our left brain hemisphere that decodes information from the right eye. Why does this matter? Because it is our left brain which houses our language centres.

Does that mean that language dictates how we categorise colours? Not quite. If you want to find out how humans categorise colours in the absence of language, then you have to ask humans who haven't yet developed language. In other words, you have to ask babies, which is by no means straightforward. Fortunately, by applying some funky technology, we can track where babies are looking at images on a screen and how much attention they pay to different stimuli. Based on such studies, we know that babies are already categorising colours before language has had a chance to intervene. What's more, they seem to categorise them more effectively with their right brains. As we age, language and the left hemisphere take over, but by then perhaps the groundwork has already been laid. As so often, the truth of whether colour perception comes from language or is hard-wired lies between the two extremes; both are important.

*

Widening the question of colour perception to include other species both reveals the diversity that exists in this respect and gives us a sense of where our own colour vision comes from. For instance, when a dog or a cat sees the world, its perception of colour is far more muted than ours. They don't see red colours or vibrant greens, but instead their spectrum is limited to two main colours, yellow and blue. We can't see the world through these animals' eyes, but based on simulations we know that what we see as red, they see as a kind of dull yellow, while our green is to them a kind of drab buff colour. Like the majority of mammals, they're dichromatic, meaning they have two different types of cones in their eye, compared to our three. Usually this involves a green cone and a blue one, meaning that their colour vision has some similarity to people with red–green colour blindness. In other species of mammals, the green cone is paired with an ultraviolet one, which allows them to see over a different, but still reduced, range of colours compared to us.

The answer to why it is that most mammals have poor colour vision when compared to other vertebrates like birds or even fish might be linked to a long-ago event that came uncomfortably close to wiping out life on this planet. Small fragments of space debris frequently collide with Earth. Most burn up in the outer atmosphere, leaving trails as they disintegrate that we know as shooting stars. Relatively few objects are large enough to make it through, and the larger they are, the rarer they are. Sixty-six million years ago, a colossal asteroid hit what is now the Yucatan Peninsula of Mexico. Based on geological evidence, it would have been around 15 kilometres across – around the size of the island of Jersey – and travelling at perhaps 20 kilometres per second when it impacted. The immediate effects were devastating, not just in the form of a crater 150 kilometres across and 20 deep, but in the way that it triggered a worldwide ecological disaster. The dust created by the impact is thought to have blocked out the sun for a decade or more, plunging the Earth into a dark, icy winter. It wiped out three quarters of life on the planet, killing all the larger land animals and most famously of all, the dinosaurs.

Our mammal lineage had inauspicious origins. The tiny, shrew-like creatures that first appeared in the fossil record something like 200 million years ago certainly offered little challenge to the established zoological order. In a world utterly dominated by reptiles, these early mammals were confined to the margins of existence, emerging at night and trying hard not to be a dinosaur's dinner. It took an event as extraordinary as the asteroid strike to force a shake-up of life on this planet. In the aftermath of the impact, the key to survival was to be small, and to be able to subsist on meagre rations. Emerging at the other side of the disaster, our miniature insectivorous forebears discovered they had inherited the Earth. A secretive, nocturnal life spent burrowing in the dark out of harm's way tends to place a limit on the value of vision. It's for this reason that even after the mammals filled the gaps left behind by the dinosaurs, their sense of vision, and in particular the relatively poor colour perception, is the legacy of more than 100 million years of evolution in the night.

Even now, the retinas of most mammals are dominated by rods rather than cones. Since rods are more sensitive to light, the upshot is that their night vison is far superior to ours. When we compare the eyes of modern mammals with those of other surviving lineages from the age of the dinosaurs, such as birds and lizards, we find that they tend to have most in common with nocturnal species of those animals. Additionally, most mammals rely on their non-visual senses to a greater extent than us. For instance, many have long snouts, tipped with a damp nose, that they use to sniff the world and collect detailed chemical information. Humans, alongside other apes, and some monkeys, are outliers among mammals in our ability to perceive colours. At some point in the distant past, gene duplication led to an ancestral primate gaining an additional, third photoreceptor and it was this that bestowed us with what is, at least compared to other mammals, our excellent colour vision. That said, even though we have three different cones, the sensitivities of our red and green cones largely overlap. This bunching up means that we don't quite get an even coverage of the spectrum. More specifically, we're extremely good at picking out subtle differences in reds and greens, as well as the colours between them. This provided a distinct advantage to animals who relied on finding fruits, many of which change from green to orange or red as they ripen. We might not rely on this quite so much in the modern world, but it's a reminder of how our senses carry the evolutionary baggage of the past and how, in a way, we look at colours with eyes that were optimised to the needs of our ancient ancestors.

We're not all equal in our ability to see colours. Something like one in twelve men experience red–green colour blindness, though it's comparatively rare among women. The reason for this discrepancy is that some of the genes involved are located on the X chromosome. Since women have two copies of this chromosome, they have a backup copy, making it vastly less likely that they will be affected by the condition. As well as being less likely to experience colour blindness, there's some evidence to suggest that women have a greater ability to distinguish between closely matched colours. As with many such variations between the sexes, there's no shortage of explanations as to why this might be. From an evolutionary perspective, it might be concerned with the role of women in early human society as gatherers of fruits and berries. Alternatively, it might come down to the debate about the role of language in perception, since women tend to have a broader repertoire of words to describe colours. A more biological explanation relates back to genetics. A slight shift in the genetic code for the red cones, combined with the fact that women have double the number of X chromosomes compared to men, means that some women may have different variants of the red cones in their retina. Ultimately, this would allow these women a small but significant improvement in their ability to distinguish colours, especially when discerning shades of reds and greens. Regardless of gender, we all lose a little of the brilliance of our colour vision as we age. A slight yellowing of the lens or cornea means that it becomes harder to distinguish subtle differences in blues and purples, and to tell yellow from green, especially when the colours are muted. The take-home message from all of this is that if a woman tells you what colour something is, especially if she's young, she's likely to be right.

While we might be among the best mammals when it comes to colour vision, among other animals, we're no better than average. Most fish have colour vision that's at least the equal of ours, while birds which have four types of colour receptors to our three, vastly exceed us. In particular, birds, like many other animals including bees and butterflies, can see ultraviolet light. When we use technology to gain an idea of what this means in terms of how they see each other and their surroundings, a whole new world opens up to us.* Seen via tech gizmos, flowers reveal hitherto unsuspected

---

* Under normal circumstances, we humans are unable to detect ultraviolet.

markings on their petals that act to draw pollinators in, while the plumage of birds such as starlings or crows is transformed. For us, a starling is a dull, mottled brown. To another starling, it's a riot of vivid purples, greens and blues. Reindeer, like some other polar animals, can also see in the ultraviolet range. To their benefit, it makes the lichen that they feed on stand out against the tundra. It has another effect too: urine appears to glow under UV. For the reindeer, this helps them see where the rest of the herd has got to and which trees a wolf has been peeing on. Meanwhile, birds of prey, which can also see in UV, can follow dribbles of rodent urine to locate their burrows. For many of these animals that signal to each other using this visual ability, it amounts to a secret communication channel.

At the other end of the visible spectrum, the ability to perceive infrared is less common among animals, although we're starting to recognise that it's more widespread than was once thought. The problem with infrared is that it's emitted by warm objects, which means that animals that generate their own body heat, such as birds and mammals, are out of the game. An honourable mention must go to vampire bats, though. Although they can't see infrared, sensors on their noses allow them to home in on a victim using its body heat. To prevent interference caused by their own bodies, their noses have special anatomical functions that keep them relatively cold. Another blood-sucker, the mosquito, also uses body heat to find a meal. Much as mozzies are universally loathed, we have to give them a certain grudging respect for the precision of their sensory targeting. They detect us by picking up on the carbon dioxide in our breath, before switching to infrared sensing as they approach, which allows them to find small areas of warm exposed skin. Rattlesnakes, pythons and boas also use

---

But some people who have had their lens removed during cataract surgery have described the ability to see the petal patterns used by bees, or the light emitted by counterfeit money detectors.

infrared sensing to find their mammal prey, using the heat that these animals produce as a means to plot their downfall. But it's arguably among fish and frogs, at the damper end of the vertebrate line, that infrared vision is most impressively demonstrated. The chromophores in their eyes are capable of undergoing a chemical reconfiguration that shifts their vision toward longer wavelengths, pushing their visual range into infrared. The effect of this is to equip them with the biological equivalent of night-vision goggles, allowing them to navigate murky water.

While all this spectrum-shifting is impressive, we need to look beyond the vertebrates to find the animal that arguably has the best eyes in the business. Mantis shrimps are crustaceans, but if the name 'shrimp' conjures the image of a diminutive crea-ture, think again. A few years ago, while snorkelling in the Great Barrier Reef, I saw a creature peering at me from the safety of its burrow. It was about the size of a banana and fronted with a colourful mass of sophisticated, sensory hardware. Though it is recognisably a crustacean, its peculiar appearance and vivid livery suggest that one of its grandparents could conceivably have been some kind of mythical Chinese dragon. A proliferation of twitching antennae collects chemical information from the water flowing past. Flattened appendages of yellow and green, used by the shrimp to communicate, wave at either side of its head. It's an extraordinary-looking animal and I was thrilled to see it, though I was careful not to get too close. Mantis shrimps are one of the reef's most deadly hunters. They carry a pair of club-like limbs braced like the arms of a boxer taking guard, just below their head. When they find a crab, or an unwary human hand, they unleash a punch of such extreme speed and power that the water in front of their accelerating shrimp fist vaporises into cavitation bubbles; if transferred to an aquarium that disappoints for some reason, they can express their displeasure by smashing the glass to effect a dramatic exit.

Impressive as this punch power is, it's the mantis shrimp's

eyes that offer most to science. Like everything else about this creature, they're colourful – large and reddish-purple and borne above the head upon turquoise stalks. Those are the colours as they appeared to me, but how they look to another mantis shrimp is an open question. While we have three colour receptors in our eyes, mantis shrimps have twelve or more, expanding the possibilities of colour vision to encompass all that we can see and more. In common with some other animals, the mantis shrimp can see ultraviolet. Yet that's not the most remarkable aspect of its supersight. Nor is it their ability to look two ways at once because of the independent movement of each eye. It's not even their facility for achieving depth perception separately in each eye. What really sets mantis shrimp eyes apart is their ability to see polarised light.

Everything that we humans see is the result of two properties of light that we think of as colour and brightness. There is a third property of light to which we are just about blind: polarised light. As light bounces around in the environment before reaching our eyes, its wave patterns get mixed up and vibrate in different directions, they're unpolarised. At other times, light hits a surface like a body of water and reflects from it with the waves now entrained and all oscillating in the same direction. This is polarised light, and we're familiar with it largely through the use of specialised sunglasses that filter out the glare. Animals that can see polarised light can use it as a kind of compass to find their way around, they can use it to improve visual contrast and thus detect things that would otherwise be hidden, and they can even use it as a secret channel of communication.

So far, so good. But why does this matter to us? Transforming a black and white image to colour allows us to see an extra level of information. Similarly, including a third property of light – its polarisation – further increases that information. Skin cancer, for example, can be hard to detect with the human eye, especially in its early stages. Using a sensor that allows us to see polarised light, however, it stands out like a beacon, allowing rapid diagnosis.

While mantis shrimps are one of many animals to be able to see polarised light, they are the only ones who can see it in all its different forms. What's more, their complex visual system is served by a pair of eyes whose internal structure is brilliantly engineered to distil the information before passing it on to its comparatively simple brain. Moreover, the streamlined arrangement of their eyes provides a blueprint that's being used to develop compact diagnostic tools that have the potential to save lives, as well as having applications in emerging technologies such as driverless cars and computer imaging.

Polarised light is crucial to all manner of animals. Bees, for example, communicate with each other using the famous waggle dance, describing in this dance the distance and direction from the hive that other foragers should travel in order to find a particularly succulent bloom. The directional part of this is done in relation to the position of the sun, but what if it's a cloudy day? No matter: if you can see polarised light, as bees can, then it's a simple matter to determine where the sun is. You might imagine that any connection between honeybees and seafaring Vikings would be tenuous at best, but both depend (or in the case of the Norsemen, depended) on highly accurate navigation and both face the challenge of what to do when you can't see the sun to take bearings from it. Bees, as we've seen, have an inbuilt solution to this on their way to a flower. The Vikings travelled on journeys of thousands of kilometres to Greenland and North America and had to improvise a solution to getting their bearings in open seas when the skies were overcast. They used a kind of crystal, referred to as a sunstone and formed of calcite, which splits noisy polarised light into two beams. By putting a dot on the top of the sunstone and then peering through the opposite side, the viewer could see two dots, caused by the separation of the polarised light. A Viking navigator on board his longship could then tilt the sunstone until he saw two equally intense spots. When this happens, the upper face of the sunstone is pointing directly at the hidden sun. By these means, the Vikings kept track of the sun, allowing their pillaging missions to be conducted with maximum efficiency.

*

The basic reason for a mantis shrimp's, or even a Viking's, use of polarised light is the same as that which underpins any of the senses. It's to do with the collection of information about our surrounds, the better to understand them. But though the senses evolved to provide intel, their importance is far greater than mere data input. They profoundly shape the way that we think and feel about the world. We can recognise this in the soothing multisensory experiences of spending time in nature, for example, or the sensual excitement of a first date. And sometimes we can strip it back to the profound influence of a single stimulus. One of the most famous examples of this was a series of experiments reported by Alexander Schauss in the late 1970s concerning the powerful effect of the colour pink. It had previously been reported that mice kept under pink lights seemed calmer and grew more quickly. Could it work with people? To test this, the experimenters held a bright pink card in front of a subject's face and then tested their strength. The results were extraordinary: of 153 participants, all bar two showed a dramatic short-term reduction in strength. A blue colour presented in the same way reversed the effect. Schauss's bright pink colour, named Baker-Miller pink in honour of two of his co-researchers, seemed to have near miraculous properties. Encouraged by Schauss, prison authorities leapt on the idea. Here was a low-cost solution to the perennial problem of violence between inmates. Springing into action, they painted entire cells a lurid pink. The idea spread to other realms when canny sports coaches tried to augment home advantage by having the visitors' changing rooms redecorated entirely in pink, right down to the urinals. After the initial excitement, however, holes started to appear in the arguments. Schauss failed to replicate his results while others found little or no effect. The idea might have died there had it not been for a Swiss researcher, Daniela Späth. The problem, she reasoned, wasn't the concept but the tone of pink used. The pink used by Schauss was a fairly vivid affair, whereas Späth opted for a pastel shade that she calls 'cool down pink', which has now been applied to prisons across Switzerland; the evidence seems to point to it providing a calming effect on inmates.

Some of the Swiss prisoners chafed at the new colour of their cells, describing them as being like little girls' bedrooms, but this association between pink and womanhood is a relatively new phenomenon. Prior to that time, male and female infants were most commonly dressed in white, or if one sex was to be dressed in pink, it tended to be boys. Blue, supposedly a daintier, prettier colour, was preferred for girls. As recently as 1914, an American newspaper, the *Sunday Sentinel*, told its readers 'If you like the colour note on the little one's garments, use pink for the boy and blue for the girl, if you are a follower of convention.'

Changes were already afoot, however. Throughout the nineteenth century and into the twentieth, it became more of a fashion for high society men to dress in more sombre, darker colours than their dandyish forebears, while women were offered a much broader palette. Gradually, pink became a colour that was worn almost exclusively by women, a colour that no 'real' man would elect to be seen in. It was perhaps to add insult to injury that the Nazis forced gay men to wear an outward sign of their sexual orientation in the shape of a pink triangle on their clothing. The gender designation of pink was amplified as the consumer society boomed in the mid-decades of the twentieth century. It's so entrenched now that in tests, girls as young as two choose pink items from a selection of coloured objects at a far higher frequency than might be expected by chance, while boys of the same age eschew it. This gender-based bias is used by marketers to target female consumers, often using pink branding, and charge them on average an extra 7 per cent for often identical items, a rort colloquially known as the pink tax. Though there is an effort to reclaim pink, and to destigmatise it, the colour carries meaning.

This phenomenon of ascribing meaning to colour seems at first to be strange. If you were to ask a visitor from another planet what the colour pink meant, what would they say? There's nothing about the colour itself that gives any form of clue. Instead, the Western association of femininity with pink is entirely subjective and culturally determined – there's no such association in Chinese culture, for example. Colour doesn't have physical properties in quite the same way as, say, the weight of an object or the temperature of a sunny day. Consequently, there are differences across cultures in how we ascribe meaning to colours. For instance, in many Western cultures, purity is symbolised by white, while in India, blue performs the same role. In parts of East Asia, white is the colour of mourning, Iranians favour blue to pay their respects and black signifies solemnity in the West. Even within a single culture, colours can have multiple meanings. Green is associated with nature in most parts of the world, but it's also the colour of jealousy in English, as in Shakespeare's green-eyed monster and it's an unlucky colour in parts of Indonesia. In China, if a man is said to be wearing a green hat, it means his partner has cheated on him. Imagine a St Patrick's Day parade seen from that perspective. Yellow, meanwhile, signifies spring in many parts, yet it's also associated with jealousy, betrayal and cowardice in parts of Europe. The French showed their contempt for criminals and traitors by painting the doors of such people in yellow. In Japan, it conveys the exact opposite in meaning, where the wearing of a yellow chrysanthemum is a badge of courage and a gesture of respect to the Emperor. It is similarly highly regarded in many African countries, while in China, if a picture is described as yellow, it means that it's pornographic.

Yet though colours may imply different things to different

cultures, the fact remains that each attach significance to them. As cultural creatures, we can't escape this. In the relentless flood of incoming data from our senses, the brain filters the information. If an emotion is attached to some element of our sensory experience, we tend to increase the attention we pay it. It could be a negative association like the hiss of a snake, or a positive emotion, like the smell of a favourite meal being cooked. Marketing departments around the world understand this fact all too well; a key part of 'experiential marketing' is using a colour on a product that conveys a message about it. Although we like to think of ourselves as complex and rational, research suggests that up to 90 per cent of our assessment of some products is based solely on colour. Colours thus convey messages, helping brands to generate a kind of tribal loyalty. Steve Jobs picked white for Apple both because it suggests purity and differentiated his fledgling company from other tech giants, who tended to use grey or silver. Cadbury picked royal purple supposedly as a tribute to Queen Victoria and grew so attached to the colour that they fought Nestlé for the right to copyright it in 2008. Advertisers not only have to differentiate their product using colour, they also have to think about how to convey the idea that the product is going to do what the consumer wants. For instance, they might use the colour red to symbolise how their product can help to stop a problem, or colours such as blue that indicate something more positive. Toothpaste manufacturers often represent cavity prevention in red, and whitening properties in blue.

This latter example hints at the idea that some colours evoke a more universal understanding amid the swirl of subjective colour associations. Notwithstanding the fierce debate in this area of research, it does seem that certain elements to our responses to colours are embedded more deeply than others. One such is the broad appeal of the colour blue. A 2015 poll asked people in ten different countries spanning four continents to name their favourite colour and blue came out a clear winner in all ten. Moreover, the preference for blue was independent of race, gender and age. Why would this be? One reason may be its association with clear skies, rivers, lakes and seas and hence with purity. People like blue, perhaps because it also seems to calm us. By contrast, red puts us on alert. It's a commanding colour – think of stop signs and red traffic lights – and it can also signify danger. Though it's often difficult to separate out learned, cultural responses from those that have deeper, evolutionary roots, one potential approach is to examine the responses of our close primate relatives to colours. A study carried out on wild rhesus macaques investigated their willingness to accept food from people based on the colour of clothing that the donors wore. The experimenters approached the monkeys wearing T-shirts and caps that were either blue, green or red and placed a slice of apple on a tray before taking a pace back to allow the monkeys some space. While the monkeys were thrilled to get a free snack, they were far more reluctant to accept it when the donor was dressed in red than when they wore green or blue, which suggests that they, too, associate red with some level of danger.

There have also been suggestions that the colour red lends an advantage in sport. Certainly Bill Shankly thought so. One of the most successful managers in the history of club football, Shankly took control of Liverpool FC when they languished in the second tier of English football and transformed them into champions. Early in his term, he switched the team's colours from red and white to all red and his reasons were rooted in the psychology of red. 'We switched to all red and it was fantastic,' he recalled. 'Tonight I went out onto Anfield and for the first time there was a glow like a fire was burning. We wore the all-red strip for the first time. Christ, the players looked like giants. And we played like giants.'

Shankly's claim that red connoted power gains some credence from a famous 2005 study examining the success of athletes competing in combat sports at the Olympics. In one-on-one bouts, each opponent is randomly assigned a red or a blue outfit, yet those wearing red win more bouts than you'd expect by chance alone. This might be because competitors wearing red get more pumped up, or perhaps because those wearing blue see their challengers as more competitive. There's even evidence to suggest that the judges of these fights tend to be more likely to award points to those wearing red. Granted, wearing red provides only a small advantage, but in the world of elite sport fine margins matter.

However, the associations with red aren't merely about putting us on alert or making us seem more dominant. Most notably, it arouses our passions in another way: it's sexy. Why this might be isn't clear, but flushed skin can be a subtle indicator of sexual arousal in lots of primates, including baboons and humans. Additionally, women's faces become fractionally more red at the stage in the menstrual cycle when they are at peak fertility – a trait that we share with some other primates. This facial glow contributes to increasing women's attractiveness to men at this time, based on tests where men studied and rated photographs. Rather more surprisingly, according to one study, women are three times as likely to wear red when they are at peak fertility. Perhaps as a consequence of this, heterosexual men tend to find women wearing red, or in red environments, dramatically more attractive. None of this is necessarily deliberate behaviour on the part of either sex, rather it's something buried deep in our subconscious that influences our actions and responses.

Red seems to motivate and provoke us more than any other colour, but the ways in which it does so aren't straightforward. It primes us for competition, it arouses our passions, and it also works more generally as a stimulant. Even as it does so, however, it can affect our performance in tasks that require thinking and problem solving. There's evidence to suggest that it puts us on edge and limits our ability to think creatively. Our responses to red also depend on the context. While we love blue as adults, as infants we prefer red. This is interesting because it suggests that we're into red largely before cultural influences take hold. What's also interesting is that even at the tender age of one, the yen for red depends on the context. Happy little one-year-olds can't get enough of red but upset them with a picture of an angry face and this preference disappears – they switch camps and pick another colour instead.

Blue, by contrast, seems to work in the opposite direction, albeit with less marked effects. If a pharmaceutical company is offering a stimulant, we tend to be more convinced by the product if the pill, or the package, is red, while if it's an antidepressant, blue works better. And if you've ever wondered why fast-food restaurants seem to choose red for their logos, their interior design and even their cups, it's partly because the colour spurs our appetites and partly because it tends to encourage spontaneous purchases. Blue works the other way round, seeming to suppress the appetite and making us more considered in our actions, which is why you don't see so many blue-branded burger joints – they've gone out of business. A word of caution, though. Important as colour is, it's only one part of an array of inputs that mesh into our overall sensory experience. Perhaps it's for that reason that there is all manner of contradictory findings in the annals of colour research.

*

If colour plays an important, if subliminal, role in shaping our behaviour, what about other visual cues? Judging by the proliferation of attractive faces that stare out from the covers of glossy magazines at the newsstand and from billboards, it's clear that marketing departments think they've found the hook to draw us in. Our assessment of a person's attractiveness is obviously something that influences our behaviour towards them when we're looking for a relationship but numerous studies suggest that our esteem for beauty extends far beyond the context of dating. Nothing exemplifies this more than a controversial program in the mid-twentieth century that drew a direct link between people's appearances and their moral character. In the immediate aftermath of the Second World War, justice departments across North America began casting around for innovative solutions to rocketing crime rates. One man, Dr Edward Lewison, a leading Canadian plastic surgeon, thought he might have an answer. Studies showed that prisoners were far more likely than the general population to have facial deformities. He reasoned that if he were to bring his scalpel to bear on the problem, he might be able to change those prisoners' prospects.

The initial results were encouraging. Those who submitted to Lewison's interventions seemed to recover their self-esteem. 'The beneficial psychological changes in these inmates were observed almost immediately,' he noted. 'There was a marked inclination to co-operate with those in authority and to participate in prison activities. Formerly hostile and incorrigible individuals became polite and gracious in their manner.'

Across two decades, Lewison offered his services, pro bono, to 450 prisoners, performing procedures including nose jobs, ear reconstructions and chin lifts. The crucial metric, so far as the prison authorities were concerned, was how these prisoners fared in the outside world, and the results were clear: the recidivism rate of the newly prettified cons was only a little over half that of their less visually attractive comrades. This on its own was not enough to convince the doubters, who pointed out that Lewison handpicked the beneficiaries of his program. Spurred by this, Lewison approached a further 200 inmates, only operating on half of them – now he had his control group. The outcome was largely the same: those who had received the surgery were far less likely to reoffend. Yet human nature is too complex to condense into a single trait, in this case, appearance. Critics of the scheme suggested that it may simply have been the attention that Lewison paid those he operated on that brought about the change in their behaviour. Others suggested that the felons were using their appearance as an excuse, and that better results could be achieved by providing training and counselling. Eventually, Lewison's scheme fizzled out, mired in controversy.

Despite the contentious nature of this kind of research, we know that the prospects of attractive people and those of unattractive people are not equal. Our judgement of others' appearances starts early. By the age of six months, babies already show a preference for looking at more attractive visages and for the rest of our lives thereafter we're suckers for a pretty face. It even has a name: the 'beauty is good' phenomenon. Relative to less attractive people, we arbitrarily ascribe qualities such as a winsome personality and excellent character to good-looking people. They make such an impression that we remember them for longer and in more detail. Studies examining the earnings of people according to their looks consistently show that there's a financial benefit to being gorgeous, even when other factors are controlled for. In fact, differential treatment begins long before we start earning. One study asked teachers to make a recommendation on a report concerning a fictional eight-year-old child with low IQ scores and poor grades. Accompanying the report was a photo of either an attractive or a less prepossessing child. It'll perhaps come as no surprise to learn that the unattractive child was judged more harshly – the assessors recommended that the homely child be sent to a class for 'retarded' pupils (as they were then known).

But what is beauty? We might reasonably say that it's hard to define, the kind of thing we recognise when we see. Given this frustratingly vague quality, it's not surprising to learn that it's something philosophers have turned their attention to for centuries in an attempt to bring clarity. Some, including David Hume and Immanuel Kant, held that beauty resides in the eye of the beholder and is underwritten by feelings and emotions. In so positioning themselves, they were running against an older argument, posited by the ancient Greek philosophers, Plato and Aristotle, who – albeit in different ways – contended that beauty

54

is an objective quality that exists independently of any one individual's assessment. We each might have an opinion on which side is closer to the truth, but to leave it at that isn't particularly satisfying. Though we might never quite see eye to eye on the question of what is and what isn't beautiful, there is clearly a lot of overlap. So why is it then that we find some things, or some people, more beautiful than others?

Beauty, as the American philosopher Denis Dutton put it, is one of the ways that evolution has of encouraging us to make good decisions regarding our survival and reproduction. It was his take that our perception of beauty has been handed down to us by our forebears and shaped by the process of natural selection. Though it has often been asserted that we learn what's attractive through being exposed to airbrushed and impossible standards of beauty exemplified by models, culture can only ever be part of the puzzle of attraction. When asked to rate people's appeal on the basis of photos, the scores given by assessors from different cultures tends to be strongly correlated. One reason for this is that faces are thought to provide cues regarding the quality of a potential mate, their genetic fitness and reproductive capabilities.

Symmetry is a major part of this; the more closely the two halves of the face are matched, the more attractive we tend to find people. In fact, not only do we find them more attractive, but raters also assess them as being more vivacious, intelligent and sociable. The differences are often barely perceptible, but they're powerful. For instance, men with more symmetrical features have more sex, with more people. Why would something as apparently trivial as symmetry matter? One suggestion is that it relates to good genes and to a related quality known as developmental stability, which essentially describes the ability of an animal (in this case, a person) to withstand the challenges of things like disease and hunger. According to this, symmetry is a sign of health. There's some evidence in support of this contention – more symmetrical men have fewer serious diseases and enjoy higher fertility – so perhaps by keeping an eye out for facial regularity, we're getting important info on a person's well-being.

The value of symmetry in persuading us of a person's attractiveness might help to explain an interesting finding that cropped up in the scientific literature late in the last century. Specifically, if you make a composite face using a large number of separate individual faces, people tend to find the composite more attractive than any of the individual faces that went into making it. This might be because the process of averaging a face irons out the quirks and asymmetries of any one face. It produces a face that closes in on the mean of a population, and we seem to like things that way – preferring faces that are in some way the prototype of a population. Composite faces might imply both genetic diversity and developmental stability, which are both excellent traits in a potential partner.

What we find attractive is also influenced to an extent by gender. While feminine traits tend to be viewed as more attractive by both men and women, what each looks for in the opposite sex isn't the same. Research suggests that heterosexual females cue into male traits such as high cheekbones, strong jawlines, more prominent brows and a slightly longer face than women typically have. Heterosexual men, by contrast, find smaller chins and noses appealing in women, alongside slightly larger distance between the eyes and a proportionally small mouth. It has to be said that research suggests that men place greater importance on attractiveness than women, but in both sexes, seeing someone we're attracted to fires up the same parts of the brain as opioid narcotics: they give us a buzz.

It's important to add that we can't distil attraction down to any single character or quality. Although there are many similarities in who we find attractive, we each have our own idiosyncrasies; part of this is to do with a rather self-aggrandising tendency in many of us to prefer faces that look like our own and part of it is due to more subtle cues in skin tone that signify health and a good diet.

But any attempt to construe some universal measure of what's hot and what's not falls foul of the manifold and esoteric ways in which we assess the allure of another person. In this context, as with so many others, visual perception is endlessly convoluted and endlessly fascinating.

Our interest in beauty and aesthetics emerges from our sensory skew towards sight. For good or ill, vision dominates the other senses, not only in terms of the attention we pay it, but also in terms of the proportion of our overall sensory apparatus that is devoted to it. Sight involves a vast number of sensory receptors – around 200 million cells – and consumes more of the brain's resources than all the rest of our senses combined. This is testament both to the complexity of vision and to the central role that it has played in the evolution of our species. It's this very complexity that gives rise to another quality of our visual lives. The challenges of understanding the ceaseless torrent of incoming information means that the brain does more work in interpreting and construing vision than it does for any other sense. Consequently, everything that we see – from an attractive face, to a landscape, to a work of art – is a construct, laden with our own personal and cultural perspectives. Perhaps it's no coincidence that we refer to our subjective beliefs and opinions as our 'views'.

# Hear, Hear!

*Sound is the vocabulary of nature.*

*— Pierre Schaeffer*

I'm nine years old and standing on a clifftop, a coat over my pyjamas, anchored by my dad against a gale that is tearing in from the sea. A storm has descended from the Arctic. The only light comes from the car's headlamps, catching a frenzied squall of biting rain and spindrift in its weak beam. I squint and turn my face into my coat's hood just to breathe against the icy wind. Above its howl, I hear the raging power of the sea, the thundering of huge waves crashing against the cliffs below. Terrified, I cling to my dad's hand. It's not just the convulsing intensity of the storm that scares me; my dad is on the lifeboat roster. Somewhere, far out at sea, there's a ship in distress and the coastguard has called out the volunteers. How could anyone pit themselves against this elemental fury and survive? I stare up at him, willing him not to go. He catches my look and half smiles, bending towards me and shouting to make himself heard: 'Someone else's dad is on that boat, lad.'

It's forty years since that storm, yet I can relive it even now. For a few minutes, we were the only people on the headland. The tempest played out in the darkness beyond the little pool of light from the car, a vicious, raw cacophony of noise, the might of nature expressed as a barrage of wild wind and saltwater. The lifeboat crew arrived, grim-faced, from the village. I was ushered into our car, curling up on the seat as my dad went off into the night with the other men down the boat slip. I turned on the radio to try and drown out the battering of the elements but there was just the fuzz of static punctuated by strange changes in pitch. Some while later my father reappeared, looking downcast. They hadn't been able to launch straight into the teeth of the storm with the waves cresting twice as high as a man. Instead, other lifeboats from up and down the coast had taken up the call, going out from the shelter of harbours. There was nothing more to do besides go back to the house, while the gale hammered at the windows and flung wreckage along the empty streets.

My memories of that night are of sound and fear. The two are often conjoined so that fear is elevated, even generated, by sound. When the director and composer John Carpenter received an early cut of his horror classic *Halloween*, he rushed to arrange a screening for a studio executive. This early cut, however, had no music and the studio executive was underwhelmed; the film didn't scare her at all. Carpenter realised he had work to do if his film was to fulfil its potential. The music that he subsequently composed was a masterful exercise in agitating filmgoers. Sudden noises that Carpenter calls 'cattle-prods' are certainly part of this, but the genius of the score is the irregular meter, an unsettling 5/4 time. The music presses and gnaws, it hurries us with its urgency and then changes pattern abruptly to unbalance us. The minor keys of the piano sound ominous, while the descending notes of the accompanying chords imply a slide into nightmarish chaos. The movie score works incredibly well by raising the stakes, drawing us in and priming us for fear. Another famous film of the time, *Jaws*, induces a similar effect, yet with only two main notes. It starts quietly and slowly, as a sinister motif, then increases in tempo and volume as the shark closes in, like a kind of musical Doppler effect that brings immediacy and mortal fear. The score isn't an adjunct to such films; it's at the heart of our emotive connection with the action.

Sound has an almost unique ability to tap into our emotions. It can evoke fear, as we've seen, but it can also conjure a host of other feelings. The tapping of rain against a window might relax us, so long as we're snug and dry indoors and have no plans to go and brave the weather. Similarly, the sound of a breeze rustling the leaves of trees can be calming and restful. The importance of sound hasn't been lost on businesses. For instance, car manufacturers pay close attention to engine sounds, tuning them to produce the most satisfactory purr for their clientele, and even ensure that the clunk that the doors make as they close appeals to the ears of the customer. Sounds that are universally loved or loathed tend to have strong and consistent associations. The beep of a car horn conjures frustrating commutes, and traffic jams. Meanwhile the pings of mobile phones are similarly provocative, cutting into our thoughts and reminding us how tethered we are by the whims of modern communications.

Loud and unexpected noises startle us. Possibly the most dramatic example of this came some years ago, when I went to scope out a new lab space with my research group. One of them, Teddy, was apparently lost in thought when another member of the team, Liss, crept up behind him, raised a rubber lobster alongside his ear and gave it a light squeeze, causing it to emit a high-pitched squeak. The effect was dramatic: Teddy collapsed to the floor like a marionette whose strings had been cut. I think Liss was expecting to make him jump rather than to cause him to disintegrate into a heap. But though the intensity of poor Teddy's reaction was a surprise, we're all prone to this. Any sudden noise, but especially those louder than around 80 to 90 decibels, roughly the same volume as a large truck going by, can induce this. It's an evolutionary trick to get us out of harm's way – we involuntarily close our eyes and often duck down, both instinctive responses to protect some of the most vulnerable parts of the body.

It's sometimes said that, across all cultures, people are born with only two basic phobias: the fear of heights and the fear of loud noises. Though this seems reasonable, we don't all respond in the same way. Why is it that some people are jumpy while others seem to be imperturbable when they hear an unexpected bang or crash? Some things certainly seem to gear the response down. For instance, oxytocin seems to suppress fear. This hormone, which is produced in the hypothalamus, regulates social behaviour, making us more co-operative and pleasant to others, as well as inducing positive feelings about ourselves. It's released when we interact affectionately with others, for instance when we're in a caring relationship. For the purposes of testing its effect on keeping us calm in stressful situations, our oxytocin levels can be temporarily boosted using a nasal spray. Following a squirt of the hormone, people still react to sudden sounds, but much less dramatically. On the other hand, being in a fearful state before hearing the noise primes us to go off like a firework. I was once on a plane, ill-advisedly watching a paranormal thriller. Midway through the film and enduring a quiet, tense scene, I was already on edge when the poltergeist, or whatever it was, suddenly appeared with a blare of noise. I let out an involuntary squeal of shock and simultane-ously my knee shot up and hit my tray table, sending a plastic cup of lemon mousse onto the lap of the guy next to me. My cry caused my neighbours to stare at me, while the gentleman wearing my pudding was very reasonably furious.

Although startling someone is a staple of pranks the world over, it does produce effects that last beyond the immediate reac-tion. People who have been momentarily panicked by the sound of a traffic accident seem to undergo a temporary transforma-tion in character in the immediate aftermath, making them roughly ten times more likely to give money to charity collectors. Strangely enough, the idea of having survived a close shave serves to make some folks nicer, for a short while at least, but I should point out that it doesn't always work this way. People with anger management problems often express an intense acoustic startle reflex. This might partly be to do with their generalised state of raised anxiety; the angrier they are, the more substantially they tend to respond to loud noises. The consequence of this is that an unpleasant surprise in the form of an acoustic shock is likely to rile them further. A feedback cycle can develop in these situations, where the anxiety makes people respond more dramatically, causing them to become angrier and more anxious. Because our behavioural reaction to these sounds is a reflex, it's largely beyond our control. It also means that we can be panicked again and again. Prolonged exposure to unpredictable, deafening and life-threatening noises experienced by soldiers and civilians in combat zones can result in chronic and very damaging conditions, such as PTSD, which may manifest for years even when the original causes are long past.

*

Despite our innate predisposition to be spooked by loud noises, we can find sounds unbearable even when they're relatively quiet. Lots of us have been kept awake by something innocuous, like a dripping tap. Once you become sensitised to the sound, it sits right at the front of your consciousness, dominating attempts to focus on anything else. I was once woken by the heavy bass of a car stereo on the road three storeys below my apartment in the early hours of a weeknight. After twenty minutes of this, and with no sign of it stopping, the last vestiges of sleep had left me and I was feeling incredibly frustrated. In the end, I got up and with apologies to the doughty chicken that had laid it, retrieved an egg from the fridge. Out on the balcony, I took careful aim and dropped my yolky missile. Gravity did the rest; the egg exploded in the middle of the windscreen and the music went off. I don't think of myself as an especially angry person, but the intrusion of the penetrating bass sent me over the edge.

Certain sounds have the ability to eat away at us, and those that do often have things in common. For one, they stand out from the background; you wouldn't hear a dripping tap during the noisier parts of the day, for instance, but at night, when everything else is quiet, it's obvious. For another, these sounds are repeated, usually fairly slow, and often irregular. And, finally, the most annoying noises are usually beyond our immediate control. Whatever the reason, the awful truth is that once the mental torch beam of our attention is on them, they become all-consuming. Little is known about exactly why this happens, and consequently, there's often little we can do to remedy the situation. Those trying to get some slumber might insulate themselves from intrusive sounds by listening to music, sounds from nature or enthralling lectures from the realm of economics, if that's their thing. However, if you do this, beware: while evidence supports the idea that these can get you to sleep faster, there are also indications that if the sounds continue as you sleep, the auditory stimulation damages your sleep patterns. Another option is to use noise-cancelling headphones. These work by using an external microphone to measure the ambient sounds around us. Armed with this information, they generate sound waves within the headset that are exactly out of sync with those ambient sounds, which has the effect of levelling the sonic playing field.

For many people, intrusive sounds are peripheral things. They're annoying in the moment, but not all-consuming. However, according to some estimates, around one in six people has a more profound response, known as misophonia, to certain noises and can find their lives blighted by it. Instead of feeling merely irritated, they get a flood of adrenaline and their hearts pound. Primal survival instincts take over and they feel an overwhelming need to escape. Common triggers include things like chewing, snoring or tapping noises, which can leave misophonics feeling like they're under attack. At the root of this is a part of the brain known as the insular cortex, an area that melds sensory experiences with sensory input. In misophonics, the insular cortex is both hyperactive and connects to other brain regions in a slightly different way to non-sufferers. Though it might seem to those of us lucky enough to be unafflicted by misophonia that it's little more than fussiness, it's real and it has a neurological basis. We also have few means to treat it effectively beyond wearing headphones, so if your significant other starts listening to podcasts while you eat, try not to take it too personally.

Although we're not all afflicted with the hatred of sounds endured by misophonics, there's a select group of noises that a large proportion of people find excruciating. Numerous polls have been conducted to identify the most commonly hated noises. We hate the whine of the dentist's drill, which is most likely because we associate it with painful procedures, and we're annoyed by barking dogs and other intrusive, disturbing noises. But the regular chart topper in this parade of audible deplorables is the sound of someone vomiting. We're also averse to the snuffling of people with a runny nose or the sound of someone chewing with their mouth open. The explanation is straightforward: bodily functions, especially those connected with the risk of disease transmission, tend to put us on alert. Responses like this are hard-wired deep in our consciousness, existing for countless generations as a means of self-preservation.

A category of other hated sounds, such as nails on a chalkboard, cutlery scraping a plate and – tellingly – children screaming, also have something in common. All of these produce intense, high-pitched sounds. Across human cultures, sonic profiles like these indicate extreme distress. Even though we know that there is no distress involved, we can't rationalise it. That's because screams and many high frequency sounds that share their characteristics are transmitted fairly directly to the amygdala, the brain's fear factory. The result is that we're predisposed to treat screams in a different way to other sounds: they bypass normal processing and put us on edge. It's not a coincidence that alarms and many other warning signals use the same sound frequencies; millions of years of evolution have left us unable to ignore them. A study examining the ability of people to identify the contexts of calls made by various animals found that we're extremely good at pinpointing those made by anguished animals. If there's a universal signal in the animal kingdom, it might just be the scream.

\*

The damaging potential of sounds is something that hasn't gone unnoticed in some quarters. The goal of harnessing its power in the form of acoustic weapons has been a focus of military research for decades. Though these might seem to belong in the realms of fiction, the kind of thing that Q might organise for James Bond, they're beginning to emerge into reality. In 2005, an American passenger ship, the *Seabourn Spirit*, was attacked by armed pirates. As the captain took evasive action, personnel on board deployed a sound cannon, which at a range of 300 m delivered a painful and – such are the strange effects of sounds on the human body – potentially nauseating experience for the would-be hijackers.\* Certainly it was enough to deter them. More recently, magnetic fields have been applied to focus beams of sound, making them into an incredibly accurate kind of sonic laser with a range of well over a kilometre. I think it would take a particular kind of person to welcome the development of a new class of weapons, though it has to be said that compared to rubber bullets or tear gas, acoustic weapons have some virtues. In particular, they don't cause the collateral damage that stray bullets or drifting gas do to people who are in the wrong place at the wrong time.

More widespread sonic deterrents can be found in examples

---

\* The vibrations caused by high sound intensities are capable of damaging our internal organs, making us feel distinctly unwell. It's been claimed that they'd make people vomit and induce spasms in their bowels with the result that they'd defecate uncontrollably. It's to be assumed that any such effects might dampen the enthusiasm of one's adversaries.

like 'The Mosquito', a device that uses a quirk of our sensory biology to target particular sections of society. As we age, the range of our hearing recedes like an ebbing tide. Past the age of twenty, we lose about one hertz per day; if the upper limit of our hearing at age twenty is initially around 20 kHz, by the time we're in our fifties, we don't hear much beyond 10 kHz. It's this phenomenon that the manufacturers of The Mosquito use. In places where teenagers loiter and engage in the kinds of special things that teenagers enjoy, the device can be used to make them suffer. There's some perfectly reasonable controversy associated with this acoustic weapon, though its advocates are quick to defend it in terms of its inability to cause lasting suffering.

Still in the realm of aural assault, a rather less subtle approach has previously been used by US forces. When the fleeing Panamanian leader Manuel Noriega holed up in the Vatican embassy in Panama City in 1989, surrounding troops played heavy metal at him through enormous, truck-mounted speakers non-stop for three days until he couldn't stand it any longer and surrendered. It's a blunt-force technique that's been used many times since, including in the 1993 siege of the Branch Davidian cult's compound in Waco, Texas, as well as against Iraqi and Afghan prisoners. Those exposed to it describe how the relentless aural battering drove them to despair. For all that our senses provide us with the means to experience the world, they also represent a chink in our armour. We can't tune out of our perceptual world simply by an effort of will, and the sensitivity of our ears can be used against us.

Although the deliberate use of acoustics against those who have incurred our wrath captures headlines, more insidious and more widespread problems are caused by sound. The modern world, especially in busy cities, is a noisy place. Road and air traffic, the sounds of people talking and going about their daily business, and even music played by stores for the apparent purpose of encouraging us to spend, all contribute to the hubbub of contemporary life. This might seem unimportant, especially when compared to other more obvious forms of pollution in our urban centres, yet it has measurable and surprisingly profound effects. Most of the research in this area has focussed on the impact on children. Comparing across different schools, an increase in background noise of 10 decibels produces an average drop in learning attainments of between 5 and 10 per cent. This is after controlling for confounding factors, such as the likelihood of children in inner-city schools being drawn from poorer backgrounds. And while it affects all pupils, the impact is greatest on those who are struggling the most with their studies. At the heart of this is the fact that noises from many different sources are apprehended by our ears, and it takes effort and energy for the brain to sort these, placing an additional load on our cognitive efforts. The saddest part of this is that it's a relatively easy fix. Basic sound insulation would work wonders, but there seems to be a lack of will among those who hold the purse strings.

As well as the nefarious purposes to which sound is being put, and the surreptitious pollution of the acoustic environment, sound can also be used positively. Various forms of sonar have been used since Leonardo da Vinci came up with the idea in the fifteenth century. Echo sounders developed as a means of seeing by sonic means were given impetus by the *Titanic* disaster in 1912. Extensive refinements of the idea have resulted in equipment that can transduce patterns of reflected sound into images, allowing us, in effect, to echolocate, just as bats and dolphins and whales have been doing for millions of years. A few years ago, I was part of an expedition with some other biologists in the Azores, investigating the behaviour of sperm whales, whose impressions of the world are gathered to a large extent by sound. As I watched, the whales lolled at the surface, circling one another, and keeping up a stream of staccato whale chatter that sounded like a rapid-firing synth drum. They'd apparently decided that the other object in the water – me – offered no threat, but it was the means by which they assessed me that was so extraordinary. The matriarch, a huge female, had approached me and, as she did so, she explored me with the exquisitely tuned sonar that these animals use to understand their environs. The sperm whale's head is, in some respects, a giant acoustic lens that focusses and directs the sounds that they make. By collecting the echoes that bounced back, she was able to form a sonic representation of me in her mind.

Whales are by no means the only creatures who can do this. Indeed, it's a skill that to some extent we have too. As a boy in Devon, Rikki Jodelko's partial sightedness didn't stop him tearing along narrow lanes on his bicycle. As he grew, however, his sight deteriorated sharply, and by his early twenties he had little more than the perception of flashing patterns of light. Faced with such a fundamental shift in the way he related to the outside world, Rikki was forced to adapt to maintain his independence. Like many visually impaired or blind people, he increased his reliance on his other senses, particularly hearing. Nowadays, Rikki can navigate using sound shadows, the result of sound waves being blocked or deflected by objects in our environment. While he insists that his sense of hearing is unexceptional, he has reconfigured the way he uses it, training his sensitivity to the soundscape, a sensitivity that few sighted people are even aware of having, and constructing a mental map of his surroundings. His skill in this respect was exemplified during a visit to Ireland to visit friends. Out on a walk in the countryside, his friends' son asked for a demonstration of his aural superpowers. The challenge was for Rikki to find the boy as he stood still and silent in a field. Using the sound shadow created by the boy's form, Rikki was able to find him easily. Although it's possible that Rikki's hearing has improved in the wake of his loss of sight, he maintains that this is a skill that we all possess, it's just that sighted people don't harness it.

*

When did you last experience complete silence? Come to think of it, is there even such a thing? Perhaps my strangest sound experience came when I went to record the narration for my last book at a studio in Sydney. The room was soundproofed with a cushioned floor and lined with thick foam rubber patterned with matchbox-sized pyramids. The purpose of this material is to absorb and diffuse sound waves, and the effect is weird. Everything sounds flat and lifeless; many people feel slightly spooked and disorientated in these places and I have to admit it was very odd. You don't notice the reverberations and small echoes that occur in normal enclosed spaces, but you feel their absence in soundproofed rooms. However, my ears were still picking up something, a faint hum that I couldn't quite place and that I couldn't explain until I read about the composer John Cage's elusive search for silence.

In 1952, Cage produced his most famous piece of work, which you might think of either a kind of self-indulgent hoax or a concept that provokes thought. The composition, entitled '4'33"', is unusual in that it requires musicians to do nothing. It's sometimes represented as four and a half minutes of silence, but there's quite a difference between quiet and silent. Cage himself discovered this when, a year earlier, he experienced an echo-free, or anechoic, chamber. Such places are used for the testing of scientific and technical equipment and represent the pinnacle of soundproofing. A particularly quiet room, such as a library, might have faint background noise that registers around 20–30 decibels. Professionally built anechoic chambers are something like ten times quieter even than this, possibly even measuring as a negative on the decibel scale. Despite this, Cage failed to find the elusive silence that he sought, realising the extent to which our own bodies produce sound. There are obvious noises, like our heartbeats or breathing, and less obvious ones like the quiet, low rush of our blood circulating* or the hum produced by spontaneous, low-level firing of our

---

* It's sometimes claimed that it's the sound of our circulatory systems that

auditory nerve. Far from being discouraged, however, Cage was inspired by this discovery. It wasn't silence that he was aiming to instil with his '4'33"', but a period of reflective calm. Rather than giving people something to listen to, he was encouraging them to listen to the often overlooked, low-level sounds around them. And though he was often pilloried for it, he considered '4'33"' to be his most important work.

*

As Cage discovered, we can never completely exclude sound from our lives. The air around us is continually agitated by pressure waves and, simple as these are, these waves are the starting point for some extraordinarily rich acoustic experiences. The act of listening lies at the heart of human language and music, and the intimate connection between sound and emotions means that sound can tap into our innermost feelings. The ability to produce and combine a near infinite variety of sounds through speech is the foundation for our social existence, allowing us to express our innermost thoughts and feelings. Language itself is sometimes described as the most fundamental of all human traits, the one attribute that more than any other has shaped the destiny of our species. Language propelled the evolution of our species, enabled us to adapt to new situations, formed the basis of cultures and, ultimately, set us on the path to developing civilisation as we know it.

Making sounds isn't unique to humans. As I write this, the banksia tree outside my window is shuddering with parrots who are engaged in a lively and ceaseless chatter. Their discussions, if

---

we hear when we lift a seashell to our ears, but this isn't accurate. If you listen to a shell in a soundproof room, you hear nothing. Instead, the shell captures ambient sounds in the environment, which then resonate within the hollow shell.

that's what they are, seem to be characterised more by their enthusiasm and raucousness than by complexity or verbal dexterity. In this way, it stands in sharp contrast to a conversation between a pair of builders who are working two doors down. Builder A has just treated Builder B, who I assume to be his junior, to his prolonged views on the relative merits of different cars and is now instructing him on the finer points of cement mixing. There's a fundamental distinction between the discourse of the parrots and that of the builders in the way that the latter are able to express a multiplicity of concepts. Human language is undoubtedly different to animal communication, but how did complex language come to evolve in our species alone?

It's a fascinating question, and one that has been the subject of debate for almost as long as language itself has existed. In 1866, the academic squabbling over the issue led one of the world's leading institutions in the field, the Société de Linguistique de Paris, to declare that enough was enough. It somewhat peevishly banned the discussion of human language evolution at its meetings, effectively killing the field for decades. During the latter part of the twentieth century, however, the topic emerged from its wilderness, reinvigorated by findings from disciplines including genetics, neuroscience and archaeology. Though there's some way to go, we're starting to piece together an answer.

Our closest relatives, apes such as chimpanzees and bonobos, are capable of communicating via a wide range of gestures and a rather smaller number of distinct calls. Impressive though these are, they're some way from the intricacies of our conversations. In fact, there are at least three crucial differences between our language and that of other apes. The first is an exquisite control over our vocal tract. The process is guided by the brain, and in particular the connection between the motor cortex, which governs our voluntary movements, and the larynx, or voice box. It's the larynx that modulates the sounds that we make, closely controlling air flow and the movements of the vocal folds. Other apes lack the neural networks that so directly connect brain and voice box, and so don't have the same degree of control. The second difference is in what Tecumseh Fitch, Professor of Cognitive Biology at the University of Vienna, describes as our proclivity to communicate. Though many animals, including our ape relatives, have the ability to interact and share information, we seem to have an inbuilt and irresistible drive to share our thoughts. You only need to watch young children engaged in ceaseless chatter to appreciate the centrality of speech to our lives. Finally, the way that our language is organised and structured allows us the unique ability to express complex ideas intelligibly. Hierarchical syntax allows us to find meaning in sequences of words, and it's the basis of human language.

Since our ancestors split from the forerunners of modern chimpanzees something like 6 million years ago, our hominid lineage has forged its own evolutionary path. In that time, we have adapted our anatomy and fine-tuned our control over it to produce ever more refined sounds. Our minds and cultures have evolved in step with language to produce an extraordinarily sophisticated facility to communicate. The question of how we made the transition from the basic language skills of our distant antecedents to our present capabilities remains difficult to resolve. It seems likely that early hominids conferred in ways similar to modern apes, that they had what's usually referred to as a proto-language. In common with these apes, it may additionally have involved a lexicon of different gestures. Even today, with our reliance on spoken and written language, gestures have their uses and are still commonly employed; I notice a few directed at me when I'm driving. During his early career as an anthropologist, David Attenborough often encountered isolated tribes in distant parts of the world. He describes how, in 1967, he was travelling in Papua New Guinea when he encountered members of the Biami people. Without any shared linguistic frame of reference, communication was only possible with gestures and facial expressions.

Nodding, shaking the head, pointing and smiling, for instance, seem to be universally understood, but what struck Attenborough in particular was the importance of our eyebrows in expressing certain emotions. In conjunction with other gestures, the eyebrows seemed to allow for mutual comprehension between the Biami and the Englishman. We quickly revert to miming and facial contortions whenever we lack a shared channel of spoken communication, and these likely offer a window into our evolutionary past.

Such approaches are, however, less than ideal for conveying detailed information*. The expansion of communication into a repertoire of vocalisations represents perhaps the single most important innovation in human history. We may never know the exact timeline of this process, but inferences based on fossil evidence seem to indicate that it began with basic symbolic communication 2 million years ago before developing into true language between 50,000 to 150,000 years ago. By mapping their thoughts onto symbolic words, our ancestors could begin to refer to people, places and objects, as well as expressing their emotions and ideas. In their earliest stages, sounds may have been used to strengthen bonds between kin, and as a means of cementing relationships that's the acoustic equivalent of the extensive grooming behaviour seen in primates. Furthermore, as it developed, it became the means by which we coordinated our efforts with those around us. At some time in the distant past, the lifestyles of our ancestors underwent a revolution. They transitioned from a diet comprising primarily of fruits and vegetation to one that increasingly incorporated meat. For a species such as ours, whose bodies lack the vital equipment typically associated with killing, such as claws, pointy teeth and explosive power, there was

---

* It's important to be clear that there is a huge difference between simple gestures and things like sign language, which, as its name suggests, is a complex form of communication.

only one way to hunt effectively, and that was by collaboration. Working together in this way requires not only nous, but a means of communication in order to marshal the efforts of a group of hunters. A basic vocabulary of shared terms provides this, setting the stage for the teamwork and innovation that is characteristic of our species.

Some years ago, I received a message from a friend that was sent via text but delivered to me after being rendered into speech by an algorithm. The content of the message was enquiring whether I fancied a game of squash, while the sound of it gave me the idea that the challenge was coming from the Terminator. Even now, with all of the developments in AI that have taken place since that message, many computerised utterances still sound wrong. It takes the endlessly complex yet beautiful medium of spoken language and flattens it, shearing the emotion from the content. Normal human conversation conveys not only words, but the sentiment behind them. Raise the tone into a higher register and we're telling the world that we're in a state of high emotion, which might be fear, or hilarity, depending on the context. Or by modulating our tone as we speak, we can convey to our listener a sense of engagement and enthusiasm. Singers of heart-wrenching refrains can convey their anguish by some subtle tricks. Adele, for instance, is masterful in this respect. In her hit song 'Someone Like You', she alternates between defiance and despond, adding subtle breaks in the voice mid-lyric that are characteristic of intense feelings when we struggle to get our words out. Doing this, she makes herself sound more fragile and vulnerable, artfully drawing us into the emotional embrace of the song.

English is often described as a non-tonal language in the sense that if we say 'banana' or any other word with a rising inflection, or with no inflection at all, or in any other creative way that you can think to say 'banana', it still just means banana. Contrast that with other languages, such as Mandarin, Cantonese or Vietnamese, where the tone used can completely change the meaning of a word. In Mandarin, depending on the tone, *ma* can mean horse or mother, *xiong mao* can mean chest hair or panda, and if you use the verb *wen*, you're either saying you want to ask a question of a person, or kiss them. While English lacks this kind of phraseological finessing, we do use different tones to emphasise particular words that can transform the meaning of a sentence. Take for instance the question: 'Did you bring a chimpanzee to the party?' Accentuate the word 'a' and the questioner seems to be implying that you should have brought several chimpanzees; enunciate the word 'party' and the suggestion is that having a chimpanzee as your plus one was a social faux pas.

Even if the language being spoken is unfamiliar, our verbal communications are encoded with readily available information that tells us about a speaker's feelings. We can be reasonably sure that the French campsite owner shouting at us is furious about something or other, just as he can perhaps intuit our confusion at his indignation. Studies that have set out to test the ability of people from different cultures to identify the feelings of a person on the basis of a recorded message show that even when their language is unfamiliar and though we don't understand the speaker's words, we can accurately identify what are sometimes referred to as the basic emotions: anger, fear, disgust, happiness, sadness and surprise. Anger is something that our auditory cortex is particularly tuned into. Regardless of the word being spoken, if it's spoken in an angry tone, it induces a dramatic increase in brain activity. This in turn primes us for a potentially threatening situation, sets our heart racing and steels us for conflict. Our ability to intuit emotion falls away somewhat when we're exposed to other vocal sentiments, such as pleasure, achievement and relief. Nevertheless, we're good at recognising feelings from voices independently of the words themselves.

In a similar way, we're also adept at matching facial expressions to emotions. This might seem so obvious that we barely stop to consider it, however it becomes more surprising when we learn

that there's quite an overlap between our own species and chimpanzees in the way that our inner state is conveyed by our faces.
This suggests that we share the way we express basic emotions,
both vocally and facially, with our closest relatives and that this
facility has ancient origins. Furthermore, our response to voices
and the feelings that they reveal is something that develops in very
early life. Before their eyes allow them to see properly, babies pick
up on the emotions underlying the voices that they hear. Even
foetuses respond to their mother's voice, and by the time they're
near full term, they show signs of relaxing within a few seconds of
hearing them read aloud. Though our linguistic abilities are hard-
wired, they take time to develop. Long before we start to acquire
words, however, we have an inbuilt facility that allows us to intuit
the emotions of the people whose voices we hear.

*

It's a short step from communicating emotion vocally to doing
so through music. Forty thousand years ago, a Stone Age person
crafted what is, to date, the oldest known musical instrument.
Made from vulture bone, and around a handspan in length, it
was carved into shape, with holes carefully drilled along its length
giving it the look of a penny whistle. The instrument was found in
2008 in southern Germany, and it pre-dates all previous evidence
of musicality in our ancestors by around 10,000 years. What we
know of the Stone Age often gives us the impression that life was
hard, dangerous and, above all, short, yet it's clear that music was
important. It cannot feed or protect us, or keep us warm, but it
does give us pleasure. It moves us, captivating our emotions in
a way that few, if any, other sensory experiences can match. We
experience an aesthetic rush, known as frisson, when we hear a
favourite tune, giving us goosebumps and a sense of reward that
stems from a squirt of dopamine in the brain's pleasure centres.
I get this feeling when I watch the clip of an orchestral flash mob

in a town square in Catalonia performing Beethoven's 'Ode to Joy'*. A solo cello player, posing as an upmarket busker, begins the slow refrain when a little girl drops a coin in the top hat at his feet. Moments later, he's joined by a second musician, and passers-by look at them quizzically. Gradually, other players enter the scene, carrying their instruments and a crowd begins to gather, at first simply out of curiosity, and then being carried by the surging inspiration of the music. By the end, there are perhaps fifty musicians and singers, encircled by hundreds of onlookers. You can see the delight in people's faces, and the excitement of the children as the crescendo builds. There's a magical sense not only of the beauty of the music, but also in the sense of people being brought together. The performance excites and inspires the audience; and even watching the video, it brings tears to my eyes.

This powerful ability of music to draw us together is perhaps what the poet Henry Longfellow was referring to when he described music as 'the universal language of mankind'. After all, most people might reasonably infer the contemplative introspection of the 'Moonlight Sonata', or the sprightly enthusiasm of Prokofiev's 'Classical Symphony', even if they aren't familiar with the melodies. Yet to meet the poet's claim of universality, music would have to transcend cultural differences and speak equally to different musical traditions. Can it? Recently, a team of researchers from Harvard put this claim to the test by asking hundreds of people, distributed across sixty countries, to identify the social contexts of short clips of music from around the world. Despite the contrasts across this range of ethnographic and musical diversity, the team found that people could reliably identify tunes as being either lullabies, ballads, healing songs or dance songs. In each of these distinct categories, the pieces shared

---

* If you haven't seen it, take a look on YouTube. It's all the more remarkable when you realise that Beethoven composed the piece after he'd lost his hearing.

common features that allowed people, regardless of their cultural background, to understand them.

This doesn't, however, mean that all cultures enjoy the same music. One of the highlights of the Yorkshire city of Bradford, where I worked after leaving school, is the amazing cuisine of the Pakistani community, who immigrated to the city in large numbers in the years following the Second World War. The food was a wonderful antidote to the drab fare that I was accustomed to, and the smells of the exotic spices were a world away from Bradford's many greasy spoon cafes. From my point of view, the only downside was the inevitable accompaniment to the food, the piped bhangra music, which to my ear sounds chaotic and discordant. Aficionados of bhangra would no doubt retort with the old proverb about people in glass houses, since my taste in music at the time centred on indie bands like The Jesus and Mary Chain and New Model Army. Musical taste is a subjective thing; we're influenced by what we hear as we grow up and we develop arbitrary preferences. According to Neil McLachlan, of the University of Melbourne, when we dislike a particular genre of music, it's often because we haven't learned its rules and its subtleties. While I'm not fond of bhangra, it's jazz – particularly modern jazz – that really upsets me. But given the number of people who do like it, it's clear that the problem lies with me, rather than with the music. During the course of his research, McLachlan discovered that training is the secret to appreciating styles that we would normally avoid. Once we educate ourselves in the structures and combinations of sounds that are characteristic of a particular kind of music, we can begin to enjoy it. This process usually happens organically, based on the music we experience in early life, but there's no reason that you can't expand your musical horizons in later life, it just takes effort.

One feature that seems to play a crucial role in shaping our enjoyment of certain musical traditions is something that the composer and philosopher Leonard Meyer described as expectation.

Once we're versed in the patterns of a genre, our brains begin to make subconscious short-term predictions, anticipating where a tune is heading. Whether we enjoy the music or not comes down, at least in part, to whether it conforms to our expectations. When it does, this is akin to the rewarding sensation that we get when we satisfy some desire. Over the course of a track, there's a constant interplay between the brain forecasting the next notes and how this matches up to what actually happens. Consequently, the extent to which we are to predict the musical motif is key to our enjoyment. If it's completely predictable – think of nursery rhymes, for example – it might fail to hold our attention. By contrast, if the tune is entirely unpredictable, it turns us off. My mystification regarding bhangra is most likely because the structure is alien to me and my brain can't work out where it's heading. Somewhere between the two extremes is a sweet spot that composers aim for, where the music is sufficiently surprising and complex to hold our interest, but not so much that we can't follow its twists and turns. When this balance is achieved, we get the buzz of gratification that emerges from an affirming hit of dopamine.

When Meyer made his suggestions about the role of expectation in musical appreciation during the 1950s, he was operating on a kind of well-informed supposition. Sixty years on, we can examine activity in the brain using imaging techniques as people listen to their favourite tunes and we can see how the nucleus accumbens, a critical part of the reward pathway, becomes more active when we hear a familiar refrain or a song that we like. What's more, we can test Meyer's contentions using these same brain-scanning approaches. From the evidence that we've gathered, it seems as though Meyer hit the nail on its head. Sounds of all kinds are transitory things, they flow past our consciousness like a river. When we listen to a story, or even to a single sentence, we need to remember what's gone before if we're to make sense of it; to do that, we use the working memory provided by the brain's frontal cortex. Straightforward as recalling recent sounds seems to us, it's something that monkeys, for instance, seem to struggle with; for them it's almost literally 'in one ear and out the other'. This in turn might help to explain why primates haven't developed the ability to produce more complex kinds of vocalisations. Humans, however, are excellent at this kind of short-term acoustic recall. In relation to music, this facility enables us to store and process aspects of melodies as we listen to them, and then to use this to shape our expectations about what comes next. Our working memory, reward system and auditory processing brain regions are intimately connected, feeding information back and forth. In this way, what we might refer to as the logic of music and our emotional connection to it are intertwined, so forming the basis for our musical tastes.

Although it's possible to gain an insight into what we like and why, music resists simple analysis. Sure, we can dissect it and gaze into the deep mathematical principles that underlie it, as Pythagoras did 2,500 years ago. We can even construct algorithms capable of creating innovative and interesting music. But in my view, when we do this, we lose something ineffable yet essential along the way. Music is, after all, art. It is intrinsically about us, for us and, most importantly, by us. This is why it resonates; it reveals something of the human condition. A Luddite such as myself might well ask: could a machine ever understand us well enough to create music that gives the same, intense connections that we have with that of our favourite artists? I'm not so sure, but like it or not, it's a frontier in music and perhaps technology might expand our horizons.

Centuries ago, Mozart experimented by rolling dice to decide how to structure musical composition. Nowadays, it's a simple matter for a computer to harvest every note of his canon, to analyse it and then use artificial intelligence to map and then imitate his style. Just like a human composer, AI can use existing works for its inspiration and so long as it can avoid overfitting, which is to say, copying too closely, the results can be surprisingly convincing. It's a long way from Wolfgang, but the band

Dadabots have been streaming non-stop death metal on YouTube for the last few years. They don't get tired because they're not human, they're an algorithm. It's an incredible achievement, particularly for its ability to generate an apparently infinite number of coherent tracks as well as for the way that it contrives music that I loathe possibly even more than modern jazz. Having said that, the Dadabots team are branching out into other formats, including rock, pop, beatbox, grunge and, inevitably, jazz. From what I've heard, it's all equally, almost viciously, awful. The AI-conjured tones of the latter sound to me like the protestations of a furious rodent in its death throes, but it is recognisably jazz and that's the impressive bit. Meanwhile, the Endel app creates personalised soundscapes, tailoring them to your state of mind and your activities. The app tracks your heart rate, where you are and what you're doing and it tailors what you hear based on this information. Extraordinary developments like this seem to suggest that human tunesmiths may soon begin to feel the metallic breath of android songwriters on their necks. For those who are resistant to the encroachment of AI into such an intensely human sphere as music, the acid test for all of this is that devised by Alan Turing. Specifically, can we distinguish tunes produced by a machine from those created by a human? To summarise the many such tests that have been conducted to date, the answer is 'probably', but the gap is closing. Then again, while an algorithm will undoubtedly arrive at the point where it can convincingly mimic a competent musician, whether AI can ascend to the heights of human musical geniuses is another question entirely.

*

Immersed in our rich acoustic environment and listening to the endless possibilities of speech and of music, it's easy to forget that hearing is based on simple physics and originally evolved not for pleasure, but for survival. What we call sound is our interpretation of vibrations in the environment. The energy of the oscillation causes excitement in air molecules next to the source. They get pushed into those further away, causing a slight compression in the air that travels as waves away from the origin, like ripples on a pond. Unlike these ripples though, sound comes in the form of longitudinal waves. They don't go up and down at a right angle to the direction of the wave's travel, but rather move more like a domino rally, pushing onward and away from the source in a series of compressions and rarefactions.

For living organisms, these waves contain information in a form that's different to that provided by light or chemicals. Having a distinct sensory channel tuned into this augments their overall perception of the world in a crucial way. A mouse on a night-time mission to gather seeds may not see other animals in the darkness, or smell them if they're downwind, but it can hear them as they move and that might well make the difference between life and death. At the same time, an owl needs only the faintest hint of a sound made by the injudicious stirrings of the mouse to pinpoint its location. Sounds not only allow these animals to spy on each other, but they also allow for deliberate communication. The grasshopper mouse of North America stands on its back legs and emits a high-pitched howl like a tiny wolf to warn away competitors, while the dawn chorus advertises birds' territorial claims and, potentially, availability for romantic entanglement.

We usually connect sound with hearing, but at a more basic level, we also feel it. Whether you stand near a speaker at a concert or watch a public firework display, the pulses of energy are picked up with the whole body. What these experiences tell us is that our senses of hearing and touch are closely related. Both spring from the same basic sense and have followed different pathways over evolutionary time to the point where we think of them as entirely distinct. Yet this distinction isn't as sharp as we might imagine. Our ears, like those of nearly all mammals, are extraordinarily sensitive, and the structures both within them and in the brain are phenomenally good at decoding the complexity of sound and unravelling its meaning. What about other organisms? As mentioned earlier, Aristotle declared that bees are deaf. We know better now, but I'm sure most people would confidently assert that plants can't hear.

I say most people, but not all. Prince Charles famously admitted popping in for a chat with the plants in the royal glasshouses and he's not alone. There's a small but dedicated band of folk who admit to doing the same, and while the plants don't offer much by way of sparkling conversation it can at least be a good stress buster for a loquacious gardener. Quite a few tests have been made of whether speaking or playing music to plants benefits them, but they so far haven't revealed anything. This isn't much of a surprise – organisms are attuned only to those stimuli that are relevant to them. That doesn't, however, mean that plants are indifferent to sound. In 2017, Monica Gagliano and colleagues at the University of Western Australia planted pea seedlings in upside-down Y-shaped pots. The above-ground parts of the plants grew in the stem of the Y, heading towards the sky while the roots grew down toward the crux, at which point the plants were forced to grow their roots either to the left or the right. Under normal circumstances, the peas respond to the choice by growing equally strongly in both directions. However, when water is on one side, the roots make a beeline for it. Normally a plant would find the water based on dampness in the soil, but Gagliano showed that moisture detection isn't the only sense that they use. She laid PVC pipes under the soil on one side and pumped water through them. Even though the pipes didn't contribute any water to the soil or affect its temperature, the peas pushed their roots through the earth in response to the siren song of the running water.

Watery gurgles aren't the only sounds that are relevant to plants. Insects can be a boon or a curse to them, depending on the bug in question. Rock cress can sense the vibrations made by caterpillars chewing leaves. The approach of these apprentice butterflies is bad news for lots of plants, but with just a little warning they can at least bolster their defences. Upon detecting the danger, rock cress plants synthesise chemicals that taste foul to the caterpillar and pack their leaves with them, making them a less attractive proposition. Impressively, plants seem to be able to distinguish between the munching noises made by leaf-eaters and other sounds, such as those made by the wind blowing through the trees, even though there can be a good deal of similarity between them. Other insects may gain the wholehearted approval of plants, such as those who rely on bees for pollination. It's a competitive world though, and the inducements that flowers offer to persuade the insects to do their business with them are a critical part of their strategy. Consequently, when evening primroses detect the buzz of a bee in their neighbourhood, they sweeten their nectar. And again, like the rock cress, the plants can tell the difference between the sound of bees and comparable sounds made by other flighty insects. None of these plants, of course, has an ear, nor any form of sensory processing organ that would be akin to our brain. What they do have are leaves and flowers that are phenomenally sensitive to vibrations. It's also interesting to reflect that plants possess some of the genes that are involved in human hearing. Though that doesn't mean they can hear, it suggests that we share elements of the sensory equipment that provides the foundations for acoustic, or rather, vibration sensitivity.

Like plants, earthworms aren't exactly dead ringers for humans. Nevertheless, they are animals and as such, they are at least a step closer to us along the evolutionary pathway. They lack ears and yet have a rudimentary ability to hear. Charles Darwin studied them long before he achieved fame with *On the Origin of Species* and returned to them like an errant lover in later life. He recognised the huge role played by these unassuming, subterranean creatures in engineering the environment and made it his mission to understand them a little better. His enquiries were extensive; for instance, he discovered that worms are no fans of cheese, but are partial to carrots – worth knowing if you plan to invite one to spend the weekend. He even worked out that around his home, at Down House near London, there were something like thirteen earthworms for every square metre of soil. However, it was in connection with his investigation of worms' senses that he really pulled out all the stops. He brought thousands of them inside, keeping them in pots on his billiard table so that he could study them close up, and then he let his imagination run riot. He flashed lanterns at them, got his daughter to shout at them and his son to play the bassoon for them. He blew smoke at them and even set off small fireworks to engage their attention. Apart from a dislike of light, the worms remained sanguine throughout.

Their sangfroid disappeared, however, when he put the pots on top of his piano and started to play. At this, the worms dashed like rabbits into their burrows, as he put it. The difference between their response on the piano and on the billiard table was vibration. Those on the piano could feel its reverberations clearly and responded accordingly. This sensitivity of worms is exploited by a band of men and women in the US known as worm growlers. Worm growling, for the uninitiated, is a process by which a wooden post is hammered into the ground and then rubbed with a piece of metal known as a rooping iron. The agitation of the post sends a medley of terrifying sounds into the ground, which in turn causes worms to shoot to the surface, whereupon they're collected for use as fishing bait. Parallel strategies are used by fisherfolk in other parts of the world, using a range of different sounds to coax worms topside. It's often asserted that the worms do this because the sound mimics the patter of rain on the ground and they're fearful of drowning. In reality, worms are in little danger of this, even when there's standing water above them. Instead, their response occurs because the sounds imitate those made by underground hunters, like moles. When burrowing predators are on the loose, the worms head to the surface, where the moles are unlikely to follow.

The ability to detect vibrations is a valuable asset to worms, and indeed to many other types of invertebrates. Many don't possess any substantial upgrade on the listening capabilities of earthworms, but some do. Honeybees, for instance, have a specialised organ that allows them to hear the buzzing of their nestmates, while male mosquitoes are sensitive to the sweet music of a female's wings in flight. The same caterpillars whose noisy eating habits alert nearby plants to their approach are themselves capable of sensing the precise frequency of the wingbeats of predatory wasps. Ingenious as all of this is, most invertebrates have comparatively rudimentary sound sensors; in most cases, there's little distinction between the sensory apparatus vital for hearing and that used for touch. This makes their hearing rather different from ours, yet if we are to understand where our acoustic sense comes from we must examine some other dissimilar animals. A good place to start is with a fish.

Anyone who has snorkelled, or even dipped their head below the surface of a swimming pool, will know that sound travels underwater. In fact, it travels over four times as fast as through air because in water the molecules are more tightly packed, so they conduct pressure waves more effectively. Consequently, hearing is an extremely valuable sense for fish, and the workings of their ears, buried inside their heads, have much in common with ours. As well as these inner ears, many fish also have a lateral line stretching the length of their bodies that is sensitive to water displacement. Both their inner ears and lateral lines are packed with specialised cells that are the basis of hearing not only in fish, but in humans. Hair cells have a bundle of up to 100 minute hairs protruding from their upper surface. Not an untidy thatch, mind you, but more of a buzzcut, whose individual hairs seem to have been clipped expertly to a very specific length, none longer than a twentieth of a millimetre. When a pressure wave enters a fish's ear, or a human ear, these hairs bend before it like stalks of grass in a breeze, triggering the firing of a message to the brain.

One of the greatest landmarks in the evolution of vertebrates occurred when, something like 390 million years ago, certain kinds of fish left the water in search of a better life. It wasn't a sudden change but a protracted one, and the earliest of these mould-breakers remained closely tied to the margins of the land, where water was easily accessible. It was a transition that ultimately gave rise to amphibians, reptiles, birds, mammals and humans. While the grass certainly looked greener on the land, everything from breathing to breeding needed to be approached in a different way, and the senses that equipped fish for perceiving their aquatic world were now less than ideal, and none was affected more than hearing.

When you're underwater, you'll be lucky to hear much that originates above the surface. Virtually all of the sound from the open air simply bounces off the surface of the water, and the same is true going from water to air. The inner ear of fish is filled with water and this is the same arrangement that they've bequeathed to all land-living vertebrates. Just like at the interface between water and air, sounds don't transfer well from the air of our outer ear to the watery inner ear. On its own, the inner ear arrangement is perfect for a fish, but pretty hopeless for any creature on land. As a result, it's long been a puzzle for biologists to work out just how hearing evolved to cope with the challenge of terrestrial life. Examining modern organisms who split their time between the wet and the dry gives us a clue. Creatures like lungfish and salamanders have been coping with this for millions of years, and it turns out that they rely on the ability to detect vibrations through their bodies as a means of hearing. For instance, many snakes augment their hearing by pressing their head against the earth, in order to pick up quivers of noise from the ground. We've moved on from this approach to hearing, yet we've retained the ability to feel sound. Rumour has it that Beethoven in his later years would hold a tuning fork between his teeth, with its other end touching his piano. The vibrations transferred through his jaw to his inner ear by a process known as bone conduction. Other parts of the skull, especially those nearest the ear, also provide an effective medium for conducting sound. Useful though this can be, particularly in the case of some types of hearing aid, for most of us it's a secondary means of hearing.

Fast forwarding in evolutionary time to the age of the mammals, we can see how the ear has blossomed into an exceptionally sophisticated sensory organ. Lodged deep beneath the head, near the tip of the jawbone, the inner ear remains the engine room for hearing, much as it has done for hundreds of millions of years. The outer ear, meanwhile, protrudes from the head. Our ears capture sound signals like a pair of satellite dishes and usher them inward to the ear canal. Having two ears is not only good for the essential symmetry of our appearance, and indispensable for holding glasses on, it also allows us to judge the direction from which a sound comes. Those coming from the side reach the nearest ear a fraction of a second before the far ear. What's more, the head shadows the sound slightly, so the signal has less intensity on the side furthest from it. Most impressive of all, we can still localise sounds in environments where sound bounces around, because the brain performs an extraordinary trick in recognising and filtering information: it uses only those noises that reached the ears first to determine where they came from. When sounds originate from above, or below, the shape of the ear fractionally changes their pitch and the brain can establish whether we should look up or down in search of the source.

The outer ear includes the ear canal and ends with the eardrum, a thin membrane about a centimetre across that vibrates in tune to whatever you're hearing. The question, though, remains: how does the ear manage to defy physics, and transfer sound from air to water? The answer can be found between the outer and inner ear in the aptly named middle ear, and it represents one of the most remarkable solutions to a problem that evolution has ever conjured: three tiny bones, the incus, malleus and stapes (in English, the anvil, hammer and stirrup), each only a few millimetres in length, connect in a chain to transfer information from the eardrum to the inner ear. This triptych of bones, known collectively as ossicles, move against each other like the levers and cogs of some madcap machine, physically passing airborne vibrations from the outer ear and into the fluid medium of the inner ear. Weird though it may seem, this odd concatenation of little bones delivers a thousand-fold improvement in transferring sound energy. Their operation is controlled by a series of equally minute muscles, which stabilise them and protect the ear from excessively loud sounds. This bizarre arrangement of Lilliputian bones and muscles improves and amplifies sound transmission and increases the range of sounds we can hear.

The weirdest thing of all is that these little bones started out performing a completely different function in the gills and other head parts of ancient fish. By the time reptiles appeared on Earth, the hammer and the anvil were built into the jaw. Only one, the stirrup, did the job of connecting the inner and outer ear, an arrangement that continues to this day in modern reptiles and birds. A little over 100 million years ago, the hammer and anvil migrated from the jaw to the middle ear in mammals. Why this happened isn't known for sure, but chances are that it's connected to something that's exclusively a mammal eating habit: chewing. The relocation of the bones was advantageous possibly because it made chewing more effective, or more likely because the separation allowed mammals to hear something other than the relentless sounds of their own mastication. Whatever the reason, the net result is that the ossicles are one major reason that mammals developed such a phenomenal sense of hearing.

Another reason for our auditory acuity can be found right next door to the ossicles. These tiny bones lever their message across from the eardrum to another membrane, the oval window. This is the grand entrance to the inner ear and, specifically, the cochlea, a cone-shaped bony tube that curls around itself into the shape of a snail's shell. The cochlea is a pea-sized worker of acoustic wonders. It's lined with hair cells that depending on where they are respond to different frequencies of sound. Those at the broadest part of the cochlea nearest the oval window respond to high-pitched sounds, while those in the tightest curl at the centre register deep, bass notes. The other bit of decoding that the hair cells do is to register just how loud a sound is, which depends on how many hair cells are triggered by the incoming vibrations.

That said, sounds can be deafening and yet not be heard by us. The song of a blue whale, for instance, is many times louder than a jet engine – yet it might as well be a mouse's whisper for all that we can hear. That's because our ears are capable of registering a specific range of frequencies, roughly from 20 hertz to 20 kilohertz; we perceive twenty cycles per second as very deep bass, while 20,000 cycles per second is a pitch so high that it'll make your dog's eyes go wide in astonishment and, indeed, when I tried playing such a sound on my computer, it made my cat hide under the sofa. Sound does exist outside of this range, we just can't hear it. Referencing our own hearing range, we describe frequencies of less than 20 Hz as infrasonic, those above 20 kHz as ultrasonic, and those in between we refer to simply as sonic. Blue whales are indisputably noisy, but their calls are in the infrasonic range and as a result, pass unnoticed by us.

There's quite a lot going on in the ultrasonic range. Many animals use these sounds to communicate, unbeknown to us. Bats, for instance, use these frequencies to locate flying prey like moths. The moths, for their part, can hear the bats' hunting cries and, if they sense that one is right on their tail, they take evasive action. If you're of an experimental frame of mind, you can try this by going to a place that moths gather at night, for instance around an electric light, and rattling a bunch of keys near them as they fly. The keys not only produce the jangling sound that we can hear, but they also give out ultrasonic sounds that are undetected by us yet alarm the moths into thinking there's a bat in close proximity. On hearing this, some of them will perform their last-ditch solution to immediate peril, which is to stop flying and plummet to the ground. Most incredible of all, I described earlier the ability of plants to register vibrations, failing to mention that plants also make sounds. It'd be a long stretch to say that plants scream, but when they're stressed or damaged, plants emit high ultrasonic sounds and they do this at a volume that's roughly equivalent of our conversation – about 65 dB. We're just not listening.

Although we're incapable of hearing the deep bass conversations of whales, or the ultrasonic squeals of hunting bats or indeed plants, the range across which we can hear is impressively broad. For this, we can thank our lengthened and convoluted cochlea. Compared to most animals, we can detect a massive spectrum of sound. Within this, however, we're particularly attuned to sounds at frequencies of between about 1 kHz and 6 kHz. It's between these two that human conversation occurs, the most salient acoustic information in our lives, and our ears have evolved over time to endow us with peak sensitivity to them. The hair cells play a big part in this, but it's a whole ear effort that tunes us in to speech. Both the dimensions of the ear canal and the workings of the ossicles are optimised to convey the specific frequencies associated with the human voice.

As well as being equipped to perceive a swathe of different pitches, we're good at distinguishing between them. As we progress along the cochlea towards the narrowest part of the coil, the hair cells respond to ever deeper sounds. When it comes to working out just how many frequencies we can distinguish, things get a little messy. This is partly because we're more discerning in some parts of the audible spectrum than others. It also matters whether a tone is pure and how loud it is, and perhaps most interesting of all, it's because hearing, like all of our senses, can be honed. Whenever we try to nail down just how capable we are of differentiating between slight gradations in any sensory stimulus, we refer to the 'just noticeable difference'. Tests to ascertain what the just noticeable difference is in respect of pitch involve asking someone whether they can tell apart two tones. Inside that person's ear, each cluster of hair cells has a peak sensitivity that's about 0.2 per cent different to the ones that come before or after them along the cochlea. But what goes on in the ear is only part of the story. In the auditory cortex, the brain is at work processing the input, and that's where the training comes in. Professional musicians can tune their instruments with astonishing precision. In the mid-range of frequencies, where our hearing is at its most acute, it's thought that some can tune by ear to within one or two cents of a reference tone. A cent is one hundredth of a semitone, so this represents incredible precision. For mere mortals such as myself, it'd be pretty good to be able to tell apart tones that differed by 10 cents; the difference lies not so much in the ear as in the brain's incredible adaptability.

Tone colour, more properly known as timbre, plays a vital role in allowing us to distinguish between sounds that have the same pitch and volume. For instance, if you hear two closely related instruments, like a guitar and a sitar, they're easy to tell apart, even when they're playing the same music. The analogy of colour with the tones that are characteristic of different instruments can be extended into the idea of a composer selecting contrasting tone colours or merging others in order to create a finished work that's rich, thoughtful and pleasing to the ear in a manner akin to that by which a painter endeavours to achieve for the eye. Timbre is what you might call the character of a tone. If you hear a pure tone, one with only a single frequency, it's smooth and unchanging, and it's also pretty bland. Try a complex tone with the same frequency and you'll hear much more going on. Its fundamental frequency – that's its lowest frequency – has a lot in common with the pure tone. But stacked on top of it are harmonics, the overtones that bring the sound alive and make it sound more interesting and charismatic. Spoken words also have this quality and it's what imbues our voices with individuality and personality. It's something that until recently was lacking in the speech of artificial intelligence, which is what made such voices seem so flat and inhuman.

Alongside the different tones that we hear, we also perceive changes in loudness, sometimes referred to as sound pressure, which corresponds to the size of sound waves* – essentially, big wave, big noise. The most familiar measurement that we use to quantify noise levels is the decibel, a measure that's based, oddly enough, on a calculation devised for the study of power loss in telegraph cables in the early days of long-distance communication†.

A reading of 0 decibels doesn't indicate no sound – it's simply a very low level of sound pressure. In practice, very few people can hear anything below 10 dB. Nonetheless, our ears are pretty impressive in that the loudest noises that we're capable of tolerating are thousands of times louder than the quietest ones; the decibel scale is logarithmic, which means that an increase of 10 dB represents a tenfold increase in sound intensity. That said, intensity doesn't relate directly to what we actually perceive; we notice volume. Our subjective assessment of sound means that we register an increase of 10 dB as roughly a doubling in volume. I say roughly because of the way that we each vary in the acuity of our hearing; one person's whisper might be another's clarion call. It also depends on the pitch. Our sensitivity to certain frequencies means that even if two sounds are exactly as loud as each other, we'll hear the one in the middle of the frequency range more clearly than another that's higher or lower in pitch.

Let's put the decibel scale into context. The average silent room isn't entirely silent. There's a little sound pressure in there, and consequently it measures around 10 dB. In such a context, you really could hear a pin drop – so long as it's on a hard floor

---

* Technically, the size of a wave is termed its amplitude.

† The bel in the word is a nod to the telecom pioneer Alexander Graham Bell, who died in 1922, shortly before this standard was adopted, and it's the reason that in the abbreviated form for decibel, dB, the B is capitalised.

– which would register about 15 dB. Under normal circumstances we talk at something like 60 dB, but in a busy restaurant, when the volume is around 70 dB, we all have to shout. Anything over 85 dB is considered to be dangerous territory for our ears, and particularly their delicate hair cells. The bad news is that these cells don't regenerate naturally, once they're gone, they're gone. Traffic noise, sirens, drills and loud music in nightclubs all fall into this range. While exposure to these doesn't mean that you instantly lose your hearing, with prolonged exposure you run the risk of damage. Since there are varying degrees of hearing loss, it's hard to put an exact number on just how many people suffer, but a reasonable estimate is around one in six adults worldwide, and the more noise there is in your life, the higher the risk.

There comes a point, however, at which sound doesn't simply cause progressive damage, and instead gets straight to the job of mangling your hearing. Deep Purple held the record for the loudest rock concert for some years when they hit 117 dB and caused members of the audience to lapse into unconsciousness. Then in the noughties the American band Manowar decided that they disliked their patrons sufficiently to hit them with almost 140 dB, something like four times as loud as a crack of thunder directly overhead. A 150 dB noise, the equivalent of standing right by a jet engine at full throttle or having a friend discharge a shotgun right beside your head, would most likely rupture your eardrum. Sounds with a volume in excess of this are, fortunately, rare; as the gauge approaches 200 dB there's a risk that the pressure waves might rupture internal organs and kill. Theoretically there's an absolute maximum volume of 194 dB because a vacuum forms between progressive wave fronts and destabilises them, resulting in a shockwave that pushes air along instead of rippling through it. Probably the loudest sound ever experienced by humans was the eruption of Krakatoa, a volcano in Indonesia, in 1883. The captain of a British ship sailing some 60 kilometres from the blast was convinced that he was witnessing the day of judgement. The explosion was heard by people almost 5,000 kilometres away. Calculations based on recordings made at stations around the globe suggest that, at the epicentre, Krakatoa's boom would have reached a staggering 310 dB.

*

While it might not seem to take much effort to register some-thing as loud as Krakatoa, hearing is not a passive process. Even when we actually start to listen to a sound, it seems automatic, simple. The brain in this regard is a little like the duck in the actor Michael Caine's phrase: it's calm on the surface, but paddling like the dickens underneath. The packets of nervous informa-tion arising from the ears have to be processed and organised. Different noises have to be interpreted and segregated so that we can focus on those most relevant to us. A network of millions of interacting neurones in the auditory cortex of the brain performs these tasks, as well as identifying the direction from where the sound originated. Despite the intricate and complex nature of all of this, the whole process is carried out astonishingly quickly. It takes around 40 or 50 milliseconds for us to perceive a pressure wave reaching our ear as sound. Just as the secret to managing is delegation, the brain processes specific sounds in different loca-tions. For most people, the right ear is the most attuned to speech and the left is dominant for music. Because of the way in which sensory information from each side of the body crosses over to the other side of the brain, the left hemisphere plays the most impor-tant role in decoding speech and the right appreciates music. Though we don't know exactly why this is so, it may allow the brain to perform simultaneous tasks more efficiently and poten-tially more quickly.

Sound is the oldest medium we have to transmit the thoughts of one human brain to enter another. It has, of course, been joined by other media, such as the written word and sign language, but it remains the preeminent force in our communication. Spoken language has played a huge role in shaping our brains, both in the way that we tune in so precisely to the pitch and volume of speech and in the development of neural tissue dedicated to decoding it. The superior temporal gyrus, a stripe of brain just above the ear, is home to Wernicke's area,* which plays a critical role in comprehending language. The advent of techniques that allow us to examine fine-scale brain activity have allowed us to explore the workings of this most mysterious of organs. Based on these, we now know that the brain prioritises speech over other sounds, deferring to words in the manner of airline staff dealing with business class passengers; vocal communications seem to be segregated from other sounds and processed separately by dedicated clusters of neurons. The ability to segregate is sometimes referred to as the cocktail party effect. It describes how when in the midst of a noisy gathering we can tune into the voice of the person we're speaking to while pushing much of the rest into the background. This selective attention increasingly requires effort as we age, and it can also be punctured if someone on the periphery of our consciousness says our name, or shouts 'help!'. What this demonstrates is that we don't completely tune other information streams out, but rather just downgrade them.

Once we're in conversation with a person, the brain attends to the business of interpreting speech in an extraordinarily specific way. Language can be broken up progressively into sentences, words and phonemes, which are usually held to be the smallest units of sound. Amazingly, the brain seems to separate it into even smaller fragments. For instance, hard consonants that we kind of

---

* Although pioneering neurologists were keen to divide up the brain into specific locations that each focus on a different task, more recent studies suggest that such apportioning is a little simplistic and that functions that were once thought to be the preserve of specific area are often carried out as a collaboration between different parts of the brain.

spit out, like d, p or t, which are known as plosives, or others that are slightly softer, such as c or s, referred to as fricatives. Vowels, too, have their own characteristics in speech. The brain allows specialised groups of neurones to focus on each part. As information comes in from each of these many different working parties, the brain puts it all together seamlessly to deliver our comprehension of what's being said.

This seamless quality stems partly from the extraordinary rapidity with which the brain interprets and assembles coherent streams of spoken words, and partly it's a trick played on us by our grey matter. Whenever there's a short interruption to something we're listening to, the brain fills in the gap. It's known as the continuity illusion, and the brain does it by using prior knowledge to predict and fill in the missing component. So effective is the brain at doing this that we're seldom conscious of it happening; indeed people under test conditions will swear blind that they heard the omitted fragment even when the illusion is revealed.

As George Bernard Shaw once said, the United States and Great Britain are two countries separated by a common language. Once, when travelling in Maryland, I stopped at a hot dog stand. After I'd placed my order, the vendor and I engaged in a bout of mutual incomprehension until the realisation dawned upon him that I wanted a hot dog. The problem came, I guess, from my flat northern English vowels; he probably heard something like *'ot dug*, whereas his eventual rendering of the same sounded to me like *hat dawg*. These issues arise for all sorts of different reasons, but the brain can be both dextrous and adaptable in this regard. Confronted by a mess of what are initially unintelligible sounds, it gets to work rearranging and sorting the puzzle until meaning coalesces from the noise. What the brain is actually doing is retuning the different areas involved in detecting phonetic features in an attempt to extract coherent information; whenever you hear a new and unfamiliar voice, your brain gets to work on the job of recalibrating to that person's tone, timbre, accent and dialect.

When we developed the written word as an adjunct to spoken language, around 5,000 years ago, we began to incorporate sight, or more accurately reading, into our communication. Our thought processes don't seem to differ dramatically according to whether the information reaches us by means of pressure waves or as photons of light. However, these are such different physical media, which engage unique, discrete pathways in the brain, that it might be imagined that the brain would treat them in separate ways. Yet when we examine what's actually going on when we read and when we listen, it seems that the brain treats the two in a way that's compellingly similar. The patterns of brain activity when we're reading correlate to an extraordinary degree with those when listening. When we either read out loud, or simply read to ourselves, the brain processes are near indistinguishable. This, most probably, is the reason that we don't draw much of a distinction between them in our day-to-day thoughts. It's the only example of its kind in nature where the brain is provided with identical information by two entirely different senses and it's an incredible testament to the human brain that it's adapted to be able to represent language independently of the sense involved.

The human ear is a brilliantly over-complicated mishmash of a solution to the need to hear. No engineer would design such a convoluted system, incorporating air and fluid and borrowing fish bones to link up the two. It converts sound waves to mechanical energy, parlays that into fluid energy, and finally transmogrifies that into electrical impulses, and as it does so, it decomposes sounds into pitch, timbre and volume. Yet for all its quirks, the ear is a phenomenally adaptable and sensitive sensory organ. As we move, or speak, muscles within the ears trim our hearing so as to dampen self-generated sounds. The acuteness of our hearing is all the more impressive for the fact that the inner hair cells, the key receptors for sound, number only around 3,500 in each ear. That may sound a lot, but compare it to vision, or smell, which each boast millions of receptors, and it's a relatively modest number.

Our ability to hear, then, is underwritten by a relatively paltry army of cells, but this is only the beginning of hearing. The evolutionary processes that equipped us with the ability to detect the endless variations of sounds in the acoustic environment also equipped us with the means not only to hear them, but to interpret them. The human brain extends this further, conceptualising sounds in the form of words and language. Indeed, the auditory brain has matched the development of the ears stride for stride to provide not only our broad perception of sounds, but the intellectual apparatus needed to achieve that most wonderful of things: to communicate with one another.

# Scents and Scents Ability

*Did you ever try to measure a smell? Can you tell whether one smell is just twice as strong as another? Can you measure the difference between one kind of smell and another? It is very obvious that we have very many different kinds of smells, all the way from the odour of violets and roses up to asafoetida. But until you can measure their likenesses and differences you can have no science of odour. If you are ambitious to found a new science, measure a smell.*

*– Alexander Graham Bell*

The air that surrounds us is a rich cocktail of chemicals. Mixed in with the gases that make up our atmosphere are innumerable molecules of other substances, released by the objects around us. We become aware of them when we inhale and they come into contact with a welcoming committee of chemical receptors deep within our noses. The molecules themselves have no intrinsic odour, yet when they interact with our receptors the result is profound – we gain the sensation of smell.

Although the molecules themselves don't smell, we'd say that the objects they emanate from do. The extent to which any substance gives off an odour depends on its volatility, or simply how readily it expels molecules into the air. The hotter something gets, the more volatile it generally becomes. That's why, when you cook something, it smells more strongly. By the same token, chilled food doesn't have the same delicious pungency as food at room temperature. This is also one reason why winter smells so much less than summer – low temperatures reduce the number of odours that are emitted.

Odour molecules that enter the nose need to be apprehended for the sense of smell to get to work. A thin layer of sticky mucus captures the molecules on their way past, rather like fly paper traps insects. Nasal mucus might be grim, but it's essential to our ability to smell. As well as protecting the delicate machinery of the nose, it reconfigures the odour molecules, breaking them up into their constituent parts. These are then escorted to the receptors, which is where the magic happens. Right at the top of our nasal cavity something like 7 centimetres from the nostril opening, sits the olfactory epithelium. About the size of a postage stamp, it's home to millions* of olfactory neurons, primed to explore the incoming chemicals.† The neurons dangle tiny hair-like structures called cilia into the mucus – they're studded with receptors, waiting to collect odour molecules.

One of the most extraordinary aspects of our sense of smell is the incredible diversity of these receptors. The human genome contains around 400 olfactory genes, each of which encodes a different type of receptor. Since we receive two copies of every gene, one from each parent, if the variants that we gain from our parents are sufficiently distinct, we might potentially end up with 800 different receptors. To put this number into context, taste uses just five or six receptor types, while our vision gets by with just two, rods and cones. This multiplicity of receptors allows us to

---

* Estimates vary on the total number, from 4 million up to 100 million.
† The flow of chemicals into the nose is determined by our breathing rate. That should mean that our perception of smell comes and goes with each cycle of breathing, or that we smell more when we breathe deeply. Amazingly, the brain adjusts for this to provide a fairly even sensation of smells.

experience a huge swathe of different odours and means that our sense of smell is unparalleled in its breadth and intricacy.

As some of my non-scientist friends sometimes say, science doesn't know everything – and in the context of our sense of smell, they're right. We don't yet know exactly how olfactory receptors recognise different molecules, but the most commonly accepted suggestion is that it's all about structure. Odour molecules come in all shapes and sizes and it's thought that the receptors in our noses use this to identify them. The usual analogy is with a lock and key: the shape of odour molecules is such that they act as keys to unlock specific receptors. The allusion gets a little wobbly when you realise that any one key can potentially open a handful of distinct locks, and a lock can be opened by a small number of different keys. Nevertheless, as far as we know, this is how our olfactory neurons work out which chemicals are in the offing.

When the receptor marries up with the right kind of molecule, it sets off a reaction that results in the neuron announcing its discovery as a pulse of electrical charge along the olfactory nerve to a pair of pea-sized structures known as the olfactory bulbs. There, the information undergoes a little reorganisation before being sent on to higher brain structures for processing. Complex though this sounds, the distance from the olfactory epithelium to the brain is short, and the nervous signals are decoded incredibly quickly. It takes around a fifth of a second from a molecule being apprehended in the nose to us consciously registering its smell. This is all the more remarkable when you consider what a smell actually is. The first cup of coffee of the day with its rich, rewarding aroma is always something to savour. You could be forgiven for thinking that there's a specific odour molecule for coffee; in fact, the smell comprises over 800 different volatile components. The various odour molecules are each detected by different receptors in the nose, and each then transmits the information to the brain. The problem for the brain is that it must unravel and decode this barrage of notifications. Fortunately, pattern detection is the brain's speciality and from this tangle of impulses it weaves the singular sensation of coffee.

Like the overwhelming majority of distinct, unique smells, coffee emerges from a chemical cocktail. Tea has a less complex aroma, yet still combines over 600 volatiles. A tomato weighs in with around 400, and even cucumber, the mildest of smells, has seventy-eight. Layered on top of all of this complexity is the fact that odours change across time. Most obviously, the chemical composition of food alters with age, going from having a delicious bouquet to a repulsive one. On a more appealing note, the molecular mix of scents given off by flowers alter as the day goes by and they switch tactics to suit insect activity patterns.

Getting the right blend of ingredients to make the perfect smell is both art and science, as well as being an endless quest for master parfumiers. Perfume has been a constant companion to human civilisation for thousands of years. The word itself comes from an Italian word meaning 'to smoke through' and describes the habit, less common nowadays, of burning incense to improve the smell of an enclosed space. Until relatively recently, perfume was used to mask the background whiff that attends the personage of anyone who doesn't wash too often, which, for most of history, meant everyone. Nonetheless, it was a mark of wealth and status, as well as lending a seductive appeal to its wearer. Nowadays our fondness for showers means that perfume has a comparatively blank olfactory canvas to work on and scents can be rather more subtle. Not that parfumiers want their product to be *too* subtle. For instance, adding a small amount of something with an unpleasant smell can be useful in developing the overall complexity of the scent. Civet oil, for instance, has a powerfully faecal smell; it could stop a charging rhinoceros in its tracks and put tears in its eyes, yet it can also bring a near magical warmth and richness to an aroma as part of a mix.

The mind-boggling intricacies of smells makes understanding our sense of smell a challenge. For our other senses, we can take a single parameter, like the wavelength of light or the size of pressure waves, and relate these in straightforward ways to our perceptions. It's much harder with smell. Alexander Graham Bell once famously challenged a graduating class in Washington DC to measure a smell, with the promise of scientific fame attached to anyone who could manage it. More than 100 years on, we still have some way to go – smell is the most challenging of our senses to measure. Though we can break an odour into its chemical constituents, it will be some time before we can accurately relate the near limitless number of combinations to specific smell sensations. In essence, each smell involves the brain calibrating and comparing the input from hundreds of different receptors. But how it renders such complexity into a meaningful sensation remains one of its best-kept secrets.

*

Despite the wonderful contributions that smell makes to our lives, it's undervalued in modern Western societies. Polls conducted in both the US and the UK reported that of our five main senses, smell was the one that people were least concerned about losing, while a study of British teenagers found that half would rather be without their sense of smell than their phone. In 2019, a survey on Twitter asked respondents to rank their senses in order of importance. By now it won't surprise you to learn that smell trailed in last. Time and again, we exalt vision and hearing to the highest places in our sensory pantheon, but why is it that we regard smell as the sensory poor relation?

It may have to do with olfaction's chequered past. For much of human history, smells were things to be wary of. The idea that sickness was borne out of noxious smells was the prominent theory in disease propagation for centuries. Clouds of pungency, known as miasmas, released from unclean dwellings, filthy streets, and even the ploughing of soil, were blamed for contaminating the body, leading to any number of maladies. A debilitating fever emerging from marshes and swamps was named after the medieval Italian for bad air: mal'aria. Terrifying epidemics that haunted the world for centuries seemed to be induced by foul, corrupted air. During the fourteenth century, the bubonic plague outbreak that came to be known as the Black Death claimed thousands of victims, condemning them to a rapid and painful end. As the sufferers deteriorated, the disease tainted them with a tell-tale, repellent stench, which seemed to confirm smell as the root cause of the illness.

Over the ages, medics used various herbs and perfumes to counter the supposedly deadly whiffs. Seventeenth-century physicians wore peculiar head-to-toe costumes, anointed with all kinds of aromatic substances, and featuring a prominent and nightmarish beak-like structure, packed with pot pourri to freshen each intake of breath. Wealthy citizens took to carrying pomanders, containers of pricey aromatics such as musk and sandalwood, around their necks and in doing so raised themselves above not only disease but the stench of the poor. The less well-to-do improvised by using a lemon for the same purpose. Noisome dwellings were set right by fumigation, while rooms were doused with strong-smelling substances like vinegar and turpentine – anything to keep at bay the dreaded miasma. The idea remained prevalent until the nineteenth century and as late as 1846, the social reformer Edwin Chadwick memorably declared 'all smell is disease'.

We only have a vague sense of how the past smelled. Odours are rather more resistant to recording in comparison with photographs, video clips and audio files. We're largely reliant on descriptions, which is perhaps no bad thing, because in the days before showers and laundries, it's likely that people and the towns they lived in would be somewhat challenging to a modern nose. However, although bad smells could never be anything other than unpleasant, we tend to perceive the smells that surround us as being normal. The process of habituation means that continuous stimuli gradually get downgraded, and the brain ignores what amounts to background noise. When we're on a long-haul flight, the air becomes stale, even with the air-purifying equipment on board. We sit in this stew of gases for hours, scarcely aware of how fetid it is until we disembark. It's a different matter for the airport staff who open the doors on arrival; they get a gust of plane gas that amounts to the olfactory equivalent of a slap in the face. Taken the other way around, astronauts who've spent time in orbit on their return to Earth describe commonplace smells, such as trees and vegetation, as heady, intense and wonderful.

*

While odours themselves were regarded with distrust, it seems like every famous man in history who ever felt moved to write about our sense of smell had some derogatory point to make (there's a notable shortage of opinions from the women of history). Most fall into one of two camps: those who regarded smell as relatively unimportant, and those who associated it with depravity. Plato considered that smell was linked to 'base urges', while others described it as degenerate and animalistic. Aristotle wrote that 'man smells poorly' and Darwin asserted that 'the sense of smell is of extremely slight service'. Freud's conclusion was that in normal development, any fascination with smells should be left behind in infancy, and he's far from alone in this judgement. We're not always comfortable with seeing people, especially adults, smell things; it seems weird to us. My own cousin, Amanda, used to luxuriate in aromas as a child, lifting all manner of things to her nose to experience their scent. The impulse was coached out of her over time by her parents, who always seemed to find it odd.

More broadly, the emphasis on observation that accompanied the Enlightenment in the seventeenth and eighteenth centuries placed a premium on vision as a means of verification; the use of the phrase 'I see' to mean 'I understand' is telling. In this context, sight is a dispassionate means of apprehending the world. Smell, meanwhile, with its fuzziness and links to emotion was relegated to the margins.

Amid the aforementioned cast of olfactory objectors, one man in particular seems to have been particularly influential in shaping our views on the human sense of smell. The nineteenth-century French surgeon and anatomist, Paul Broca, was a pioneer in the quest to understand the brain. In true mad scientist style, Broca kept hundreds of them in his Paris laboratory, preserved in jars of formalin. His fascination was with its different structures and, in particular, the frontal lobe. This area, situated right behind the forehead, is critical to all manner of functions, including consciousness, memory formation, empathy and our person-alities. Compared to other mammals, primates are blessed with especially large and complex frontal lobes, and among primates, humans emerge as being disproportionately equipped in this area. The goal in these early days of neurology was to relate function to the different parts of the brain and in this initiative Broca was extremely successful. He's best remembered for his research into the parts of the frontal lobe that govern speech production, which now bear his name: Broca's area.

Yet even while Broca was advancing his field, he was formulat-ing other ideas that remain influential despite their shortcomings. In effect, he was searching for evidence of human distinctiveness and he considered that he'd found in it his measurement of differ-ent species' brains. He contended that in humans, the olfactory bulb, the part of the brain that receives information from smell receptors, had shrunk to make space for the enlargement of the frontal lobe. Adding two and two together to make twenty-two, Broca seized upon this as proof of mankind's elevation above the other animals. The frontal lobe had 'grabbed the cerebral hegem-ony,' he crowed. 'It is no longer the sense of smell that guides the animal.' Thus, smell was a base, primitive sense, and shedding it was evidence of human superiority.

In the intervening years, many scientists have followed Broca's lead and arranged the facts to fit their theory. In reconstructing human evolution from a sensory point of view, it has been claimed that our transition to walking upright took our noses away from the ground and so away from the richest whiffs and pongs. In the process, so it's been argued, we relegated the importance of olfaction. Going back further, the migration of primate eyes to the front of the face allows excellent stereoscopic vision, compared to the side-of-the-head arrangement favoured by many other mammals, but limits the space available for the olfactory equipment. The loss of the snout in apes especially seems only to further restrict the capacity for smell. Finally, primates in general and humans in particular seem to be losing genes associated with our sense of smell. We have something like 400 working olfactory genes, but sitting in our genetic code are close to 500 olfactory pseudogenes. These are the genetic equivalents of fossils; genes that used to contribute to our sense of smell but that no longer work. In other words, we've lost over half of our smell genes across evolutionary time.

If you were to look more broadly across mammals in general, you might see a pattern. For instance, whales evolved from land-dwelling animals but millions of years of aquatic life with only brief intervals at the surface means that the ability to smell airborne chemicals isn't of much use to them. Consequently, they've lost virtually all of their working olfactory genes, leaving only the fossil pseudogenes behind. At the other end of the scale, rats, which rely extensively on their sense of smell, have both far more working olfactory genes than us and a far lower proportion of remnant pseudogenes. In other words, animals that rely on their sense of smell tend to have more functioning olfactory genes.

The idea that we're visual creatures who lack a sophisticated sense of smell has become such a powerful narrative that the tide of opinion has often seemed to carry all before it. Even today, olfaction is given vastly less attention in the field of scientific research than the other senses. Based on the number of studies published on each of the human senses, smell is accorded approximately a quarter as much consideration as hearing, while over twenty times as many papers focus on vision than on smell. It's perhaps for this reason that while we can detect gravitational waves, find water on Mars, and grow human organs, we still can't claim a complete understanding of how our noses work.

However, a new perspective is emerging that suggests that we're far from Broca's notion of us as olfactory poor relations. Though it's true that when you compare the size of the human olfactory bulb with that of our brain, it isn't as prominent as those of other animals, this overlooks something rather important. Though it may be proportionately small, in terms of overall size our olfactory bulb is larger than that boasted by a mouse and has at least as many neurons. The human brain has certainly grown over evolutionary time, but that doesn't mean that our olfactory centres have shrunk. Broca and his contemporaries were content to infer function from structure. If a particular area of the brain was disproportionately large, that meant it was correspondingly important. Though a scheme like this might yield some general clues, at best it lacks finesse. At worst, it's useless.

What about the underlying genetics? We know that with around 400 functional smell genes, we lag far behind animals. Mice and dogs, for instance, have more than twice as many olfactory genes as us, while elephants have around 2,000 olfactory genes. However, that's only part of the story. For one thing, 400 receptors is vastly greater than the number of different receptors we have for our other senses. And even acknowledging that we're inferior to some other mammals in terms of our olfactory equipment, the processing capabilities of the human brain are without equal. In other words, it's not what you've got, but what you do with it that counts.

With the more refined experimental approaches available nowadays, our perspective on human nasal prowess has undergone something of a renaissance. Back in 1927, a research paper made an offhand suggestion that we could identify 'at least 10,000 smells'. As often happens, this statistic took on a life of its own: 'at least' was dropped and generations of students learned that 10,000 was the golden number for the human olfactory palette. It took until 2014 for this to be seriously challenged, when a team of researchers led by Caroline Bushdid from Rockefeller University concluded that humans are capable of distinguishing well over a trillion different smells. A trillion! Compare that to the scope that our two supposedly dominant senses, vision and hearing, can manage: we can hear perhaps 1,000 different audible tones and identify up to 10 million different colours. Suddenly, our sense of smell no longer seems like the poor relation.

How about our sense of smell in comparison with other animals? The way that comparisons are typically made is by determining the lowest possible concentration of an odour that an animal can detect, as a measure of olfactory acuity. Humans are capable of registering the presence of certain chemicals at concentrations of less than one part per 100 billion – the equivalent of less than a drop of water in an Olympic swimming pool. Across a spectrum of biochemicals, human subjects performed surprisingly well, in some cases better than acclaimed nostril champions like mice or dogs. Of the fifteen different chemicals where humans have gone head-to-head with dogs, we've come out on top in five while dogs have prevailed in the remainder. The substances that we're most sensitive to are derivatives of fruit and flowers, while the odours that dogs tuned into are all found in the smells given out by prey animals. In both species, the sensitivity to particular odours has been tuned by evolution. The earliest humans had much in common with modern primates, particularly in regard to diet. Our ancestors were herbivores and a sense of smell that guided them towards fruit would have given them an advantage. Wolves, the forebears of dogs, tend to key in on olfactory cues that fit their carnivorous lifestyle. Comparing olfactory abilities isn't always as simple as saying which has the best sense of smell, it depends on what's being smelled.

It's important not to overreach, however. Dogs are among the best snufflers in the animal kingdom and far outstrip us by most measures. The collaboration between our two species is one of the most extraordinary and long-lasting partnerships in nature. Dogs, and their feral predecessors, not only guarded our early settlements, as they still do today, they dramatically increased the effectiveness of our hunting. Dogs are adept at picking up odours, for sniffing out a trail to find hidden quarry. As dogs zig-zag along a trail, their brain is comparing minute differences between the inputs from their two nostrils. This ability to use nostril comparisons to locate smells isn't unique to dogs. A few years ago, a group of researchers at the University of California, Berkeley, laid a trail of chocolate essence across a lawn and challenged blindfolded volunteers to channel their inner bloodhounds. It seemed a tall order, yet by crouching on their hands and knees with their noses pressed to the ground, the students managed to locate the odour by contrasting the strength of the smell in each nostril and following it along the grass. Admittedly, the human subjects didn't find the task quite so easy as a dog might have, but as they repeated the test, they got better each time.

This improvement is characteristic of both smell and taste; both can be honed and sharpened through training. It's precisely for this reason that dogs, long used as hunting companions, have become sophisticated detectors of hidden contraband or even sniffing out disease. The aforementioned bloodhound is generally reckoned to be the master at searching for evidence and escaped criminals. While the human olfactory epithelium is about the size of a large postage stamp, the bloodhound has something like forty times as many receptors in an epithelium that if flattened out would be about the same in area as an iPad. Their exploits are legendary – following trails as old as twelve days and for distances of over 100 kilometres. The extent of dogs' olfactory perception aligns them with the animal kingdom's other super-sniffers. While in the days when Paul Broca's ideas held sway, the spectrum of mammal smellers was depicted with dogs at one end and us at the other, the most recent comparative studies suggest that we have an excellent sense of smell that puts us around the middle of the mammalian pack.

\*

Alongside the new-found appreciation for the human senses of smell, we're also starting to understand the extent of the differences between us. Shortly after starting my PhD, I went with the rest of my research group on a field trip to Canada, leaving Europe for the first time. New sights and sounds were to be expected, but what I hadn't anticipated was a whole new smellscape. The rich, resinous smell of the pine woods where we worked, with their background hint of citrus, was different to that of any forest I'd experienced at home. Then there was the stunning smell of skunks, whose aroma, a blend of sulphur and burning rubber, belies their charming appearance. There were dozens of new odours, including one smell that was the olfactory equivalent of assault. It was cloyingly, nauseatingly sweet and it assailed me every time I came within a few metres of my colleagues. Eventually, I tracked the smell to its source; it turned out that Dan, a senior member of the team, was anointing himself with a hair product that made him smell like he was dousing himself with caramelised vomit. I tried to drop hints over the course of a few days to no effect until, in desperation, I took matters into my own hands. Shamefully, I waited until everyone else was distracted one evening, located the noisome tub of sludge, and hid it. Dan was perplexed the next day when he discovered it was missing but, at last, I could enjoy his company without dry heaving.

Why was I the only one among us to be laid low by the smell of this noxious product? The answer is that each of us has a unique nose, a different perspective on the world of aromas. Take any group of people and ask them to smell something – let's say a lemon. They would be likely to agree on the general label of lemony-ness but its intensity and other qualities like pleasantness of the scent will be a subjective experience. At the root of this is the extraordinary number of receptors dedicated to smell, and the genetic differences between us. Each of our 400 smell receptors is encoded by a particular gene, but aside from the genes that underpin our immune system, our olfactory genes are the most diverse in our bodies. Across the world's population there are at least 1 million variations of our 400 olfactory genes. In other words, the chances of any two people having the same battery of receptors is basically zero. In fact, it's been suggested that any two people differ in this respect by about 30 per cent. Ultimately, we each have our own exclusive experience of smell, which is sometimes referred to as our olfactory fingerprint.

Criminologists have used fingerprints to identify criminals for over a century, comparing the idiosyncratic patterns of a print against a library of other prints held in a database. Getting an accurate match depends on the number of unique characteristics, known as minutiae, on the fingerprint. Essentially, the more minutiae the better, but anything more than about forty is considered pretty infallible when diagnosing a match. The principle with an olfactory fingerprint is surprisingly similar. To take an olfactory fingerprint, someone is asked to characterise a series of odours using a list of descriptive words. The differences between people's senses of smell are such that it's potentially possible to identify every person alive by collecting their responses to thirty-four distinct aromas. A person's olfactory fingerprint is largely dependent on their genes and remains consistent over time. The point of olfactory fingerprinting isn't to catch felons; it's hoped that the technique can be used in medicine, for instance as a non-invasive method for finding donor matches for organs and bone marrow.

Given the potential trillions of different smells that we can each experience, it's perhaps no surprise that we don't have a catalogue of our responses to all of them. Nonetheless, there are some fascinating case studies that highlight the differences between us. I'm not often engaged by the contents of vending machines in public toilets, but some years ago I came across a product that I found comical and ludicrous. The product in question was a spray which promised that, when applied liberally by its male purchaser, it would cause women to find them utterly irresistible. It turned out that the magic ingredient in this concoction was androstenone, a substance that is unrivalled in its ability to inflame the passions of pigs. More specifically, it's a pheromone produced by boars that renders sows cross-eyed with lust. It doesn't have the same effect in humans, not least because there's no evidence of any kind of pheromonal communication in our species, but if a person were to injudiciously apply the spray to themselves in the vicinity of a pig farm, the resultant outpouring of love from an army of besotted sows would make for a fascinating spectacle.

But while androstenone doesn't excite us in quite the same manner as pigs, it's a useful subject of study since it provokes a diversity of responses in people. Some can't smell it at all, while a small proportion report it as having a pleasant, fairly sweet odour. A far larger number of people, however, describe it as smelling sweaty and urinous. This isn't simply a matter of personal fussiness; the responses of each of these three groups is determined by their genes. Apart from putting a twinkle in the eyes of lady pigs, androstenone causes what's known as boar taint, which makes pig meat taste disgusting to some. When our species first emerged, we carried a variant of an olfactory gene that makes androstenone smell terrible and which makes us averse to eating pork. However, as humanity expanded beyond Africa, a mutation in this gene changed our perceptions. Carriers of this variant either couldn't smell androstenone, or were indifferent to its scent. This set the scene for the domestication of pigs in the Middle East and, separately, in what is now China, and led to pork becoming a staple of the human diet in many parts of the world. Even today, the original gene variant is most common in Africa, while in Europe and Asia, the variant that is more forgiving towards piggy smells tends to hold sway. More generally, the range of different responses to a single substance stands as an illustration of the incredible diversity of human sensory perception.

\*

In addition to genes, culture plays a crucial role in sculpting our perceptions. The comparative disregard for the sense of smell in the West can clearly be seen in language. In English, almost three quarters of the words we use are based on our visual experiences, compared to less than 1 per cent that relate to smell. It's not just that we don't talk about smell, either; we also don't think of it as a valid means of appreciating others. For instance, we might refer to someone positively as being good-looking, but if we say someone smells, it's unlikely to be a compliment. If we qualify it by saying 'you smell good', we're most likely to be admiring their perfume rather than their natural odour. More generally, when we attempt to convey a smell in words, we usually have to relate it to something else, as in 'it smells like cinnamon'. Apart from words like 'musty' or 'fragrant', which are themselves not especially informative, this is about the best we can do.

Our stumbling attempts to parse aromas influences the way we think of them. The failure of English to get to grips with smell gives rise to the idea that smell can't be got to grips with. Odours seem vague and ineffable, especially in comparison to colours or sounds. Smells themselves, of course, are very often not vague at all, as anyone who has had a close encounter with a skunk will testify. Nonetheless, the lack of clear terms in our vocabulary has caused us to think of smells as being nebulous. This isn't unique to English but also applies to other European languages; smell just isn't as present in our conversations as other senses are.

At the same time, the marginalisation of smell has only accelerated as we've moved into cities and away from nature, taking up a modern life that has a limited and constrained smellscape. For one thing, air pollution is well known to stymie our sense of smell. For another, we spend ever-increasing amounts of time indoors; the average American spends more than two thirds of their waking hours indoors. The glorious mix of natural smells that have been our companions since we left the trees are now shut out, separated from our starving noses. The screens that have come to occupy ever more of our attention stimulate senses like vision and hearing, but offer no smell sensations.

The mix of our language deficit and these changes in our lifestyle explains why Westerners perform staggeringly poorly when asked to name commonplace smells under test conditions. You'd imagine that most could correctly identify coffee, but in a study employing blindfolded American undergraduates as olfactory guinea pigs, only one in four got it right first time. Orange was easier, but even then, only half of the students managed it. Something like 20 per cent of them correctly identified the smell of basil or peanuts, and very rarely was cheese identified from its odour. It might be understandable that we in the West have concluded from such results and the near absence of smell in our languages that this must represent the general human condition, but if you look beyond the West, the picture changes.

In the last few decades, the scope of research has encompassed other cultures and the results have been nothing short of astonishing. The sense of smell is accorded enormous importance by the people of many cultures – for many, odour is at the heart of their lives. The Desana people of the Amazon rainforest, an environment rich with sensory stimulus, describe themselves as *Wira*, or 'the people who smell'. From an early age they're capable of following 'wind threads' or scent trails, identifying plants and animals by their aroma and navigating the forest using a sensuous network of smells. Other tribes in the area are distinguished with a particular group odour, borne of the place where they live. It's of such importance that it guides them in their choice of partner, since spouses must have a different smell. Like the Desana, another Amazonian tribe, the Suyá, similarly categorise animals according to their smell. They also ascribe odours to different members of their communities, distinguishing between the sexes and stages of life. An individual's scent is of such importance in some cultures that they take precautions to avoid mingling odours. A person's smell is their essence and mixing yours with someone else's, such as when two people sit too close for too long, is carefully regulated.

The Ongee people of the Andaman Islands also live in a world where smell is elevated. Each season is defined by the aromas of flowers that bloom at that time, meaning that the calendar is set by the smellscape. The Ongee believe that evil spirits locate their victims by smell; when moving through the forest, they travel in single file so that their odours blend. When they greet one another, they ask 'How is your nose?' Strange though this might seem, scent is important in the greetings of many cultures. In parts of India, the most tender greeting was once to smell a person's head. Rubbing noses and sniffing one another has long been a traditional greeting of peoples from the Arctic to Polynesia, and from West Africa to the Philippines. In the Gambia, it was once common to sniff the back of a person's hand as a means of saying hello.

A Desana or an Ongee child learns about their world through sensory experience in a way that is fundamentally different to how we learn in the West. A youngster in Britain or the US might Google the image of a particular flower, whereas such identification in other cultures is a more holistic sensory experience that encompasses scent, touch and taste. The central importance of odour in certain cultures hones their sense of smell to the point where it resembles a superpower. An experiment comparing the sensitivity of smell in Europeans to that of an indigenous people from Bolivia, the Tsimane, found that the Bolivians were not only substantially better than the Europeans on average, but that a quarter of them outperformed the most sensitive of the 200 Europeans tested, even though at the time of the experiment, many of the Tsimane were suffering from colds.

The sensory experiences of different cultures, and in particular the value that many place on smell, contributes to a greater diversity of perception, and it's an experience that is enshrined in language. The Jahai people of Malaysia are hunter-gatherers who have a rich lexicon of words dedicated to odours, which not only reflects the importance of smell in their lives but also shapes their responses. In contrast to the verbal fumbling that characterises our struggles to describe odours effectively, the Jahai do so with the same consummate ease that we experience when we describe a lemon as yellow. Their ability to relate what they smell to concepts that they carry in their minds improves their ability to evaluate them and opens the door to clearer and richer perceptions of odours. Research by Asifa Majid of York University showed that not only were the Jahai faster and more accurate than Westerners in identifying smells, they also arrived at a much closer consensus. To them, a smell is clear and obvious, a shared reference. These findings completely overturn the idea that smells are nebulous sensations that defy description. It's a sense that we've long neglected in the West; with this new perspective, it's apparent just how much we're missing out on.

It's not just genes and culture that shape our ability to smell. As with all our senses, there's a decline with age. The ability to smell decreases until by eighty, when three quarters of people have major deficits in this sense – which explains why older people going out on special occasions have a habit of wearing enough scent to stun a mule. This decline is the consequence of a loss of both olfactory neurons and mucus, the essential lubricant for our sense of smell. Sadly, there's a risk that with age, other factors will come into play. The loss of the sense of smell is one of the symptoms of many diseases. Most recent, of course, is the sudden loss of smell suffered by almost half of all people who contracted

Covid-19. While the majority eventually recovered their sense of smell, it's not yet known why Covid affects us in this way.

Sex, too, affects smell. Bad news for men, but women are the sensory superstars of our species and that applies to smell, too. When a man feels he's being unfairly chastised by his lady love for a lack of hygiene, the problem may well be that he can't smell quite as effectively as she can. His own personal miasma might be lurking below the threshold of his notice while simultaneously making her eyes water. Women can not only perceive odours at lower concentrations, but that they have a richer olfactory experience more generally. Much of this difference comes from the fact that they have up to 50 per cent more neurons in the olfactory processing parts of their brains than men.

As so often, one question leads to another: why do women have a better sense of smell? It may be to do with the importance of the bond between mother and baby at birth, or it may be connected to the choice of mate. However, the most plausible explanation concerns pregnancy. Much as smell is important in rewarding us with pleasant odours, its primary function is to help us avoid toxic materials. Things that may have little or no effect on a healthy adult could be incredibly damaging to a foetus – through evolutionary time, those women whose sense of smell enabled them to avoid such toxins would have had healthier children. This might also explain why women's sense of smell is sometimes amplified during pregnancy and some become highly susceptible to certain substances. For example, one of those chemicals that makes up the smell of coffee, indole, can cause problems for coffee-loving mums-to-be. To most people, indole has the odour of bad breath or faeces. In the melee of hundreds of different aromas in coffee, hardly anyone notices it. When some women become pregnant, indole's shitty odour comes to the fore and ruins the whole experience.

You don't have to be pregnant to be a supersmeller. Somewhere between one in ten and one in fifty people is gifted in that way, and technically they're known as hyperosmics. What it is about these people that provides them with this sensory superpower is something of a mystery. They might be benefitting from a lucky throw of the genetic dice, but the most likely candidate is simply hard work. The more familiar you are with a scent, the lower your threshold for detecting it – the sense of smell improves if you put a little effort into training it. This has long been known among sommeliers and parfumiers, whose brains adapt to the demands of their sniffing. Seen under a brain scanner, people in these specialist careers have olfactory processing areas that show clear signs of restructuring and improvement. Just like bodybuilders, the longer a parfumier has been flexing their olfactory muscles, the more improvement they show. These training effects needn't be the preserve of professionals – even regular, short* bouts of exposure to unfamiliar odours can yield impressive results. One study, performed by Syrina Al Aïn and colleagues at the University of Quebec, revealed that after only six weeks, the brains of people who underwent smell training showed clear evidence of development and their olfactory performance improved. Given the importance of smell to our sense of well-being, it can't hurt to have a go.

*

As with our other senses, there are two sides to the olfactory coin. One is the ability of an individual to smell, the other is the odour that each conveys to others. Among the questions that I've wrestled with in my career is how animals recognise one another. It turns out that though there may be hundreds, or even thousands of individuals of a given species in a local area, the way that they interact isn't random. Animals, like us, are cliquey; they hang out with certain close contacts for a disproportionate amount of

---

* Just a few seconds daily seems to be enough.

time, while ignoring others. On my trip to Canada, my attention was on small fish. You might think that a creature like a stickleback wouldn't be choosy about its companions, yet it turns out that they are extremely discerning when it comes to socialising. Having studied who was hanging out with whom, we analysed the population and found that the fish in our Canadian lake were part of a typical 'small world' network. Even as they move around their habitat, they organise into clusters of close companions, while having looser affiliations with acquaintances from other clusters.

The question is, how do they know who's who? For most of the animals on the planet, smell is the guiding force in recognition and fish are no exception. They can identify each other because every individual has its own distinct personal odour, like a chemical signature. This smell identity is so important that for many species, it overrules the other senses. I once ran an experiment where I gave fish the choice of shoaling with their ilk, or with a different species altogether. With a bit of experimental trickery, I offered them a choice between an option where they could see others of their own type but smell a different species, versus one where they could see the other species but smell their own kind. What I found was that they trusted their noses – it didn't matter that the animals they could see didn't match up, what was important was that they smelled right.

Fish, like all animals, produce a bouquet of smells that tell others pretty much everything they need to know. Some animals try to mask their scent, with varying degrees of success. Which dog owner hasn't thrilled to the sight of their freshly manicured pooch luxuriating in a blissful roll in some manure? Somewhat similarly, assassin bugs glue the carcasses of their victims to their carapace. In both cases, the animals are trying to smell less like themselves and more like the things they'd hunt. Humans, meanwhile, attempt a different form of olfactory masquerade – it's estimated that we spend over $20 billion a year on deodorants and about twice that on perfumes. Ironically, we veil our own scent with aromas from other animals, often from glands situated at the unspeakable parts of a deer's, civet's or beaver's body. As well as things like musk and castoreum from these animals, perfumes often contain a hint of urinary fragrance. Horrible as it may sound, our noses seem to like it. Nonetheless, whether you're a dog, a bug or a person, there's a limit to how far you can keep your smell at bay.

Humans are no different to other animals in that we each have our own distinctive scent, which is partly to do with our metabolism and the hordes of bacteria that roam invisibly on our skin and partly to do with our genes. The fact that our genetics affects our smell provides us with a means to recognise who we're related to as well as who we might like to get to know. The major histocompatibility complex,* or MHC, is a suite of genes that play a critical role in our immunity to disease, promoting our ability to recognise invading pathogens. In addition to this work, the MHC strongly affects our chemical signature. Since MHC genes are highly polymorphic, they give each of us a distinct scent, and we can tell people apart based on this genetically determined odour. Because we share genes with our blood relatives, our kin smell similar to ourselves. Mothers can single out clothes worn by their own babies by smell alone, and babies can identify their mother from the scent of a breast pad. In a similar way, children can identify their siblings; identical twins smell so similar that tracking dogs can be gulled into following the wrong trail if a different twin steps in. All of this promotes kin recognition, ultimately allowing us to assist our closest relatives and helping us to avoid the perils of incest by making close relatives smell sexually unattractive.

The role of our scent in sexual attraction is an area that has long been discussed. Back in 1995, the Swiss researcher Claus Wedekind asked female volunteers to rate the scent of T-shirts

---

* In humans, the MHC is sometimes referred to as the HLA – the human leucocyte antigen complex.

that had been worn for two days by male students. To level the olfactory playing field, the T-shirt donors were asked not to wear any deodorant or perfume. The results were remarkable – women showed a strong preference for T-shirts worn by men that had different MHCs to themselves. There's sound biology behind this preference – it's nature's way of avoiding inbreeding. It's even been suggested that this may be why we kiss – it helps us sample a potential partner's smell and taste. While the thought of this strips the romance from the situation, it illustrates how smells can subconsciously guide our behaviour. Although it's rather unlikely that the women in Wedekind's study were envisaging a prolonged relationship based on the smell of a T-shirt, how attractive you find a person is guided to an extent by an assessment of their suitability as a mate. The point of having two parents whose genes are recombined into a child is to spread the genetic risk. So having children with a person who has a different MHC to yourself means that the offspring have a higher chance of inheriting a broad spectrum of MHC genes and, ultimately, having a more competent immune system. In other words, the women's sense of smell was guiding them toward healthier future children. Interestingly, though, if the women involved in rating T-shirts were taking oral contraceptives, their preference was reversed; they were inclined to like the smell of men with similar MHCs. This tells us first that hormones influence mate-choice decisions and second that since the pill works by essentially tricking the body into thinking it's pregnant, it might also cause women to be drawn toward their social ingroup, perhaps including nurturing relatives.

Since Wedekind's results were first published a quarter of a century ago, multiple follow-up studies have been conducted, and the results have been mixed. Perhaps we shouldn't be surprised – after all, human attraction is a complex and intricate business. The acid test for these ideas is whether rating the attractiveness of a whiffy T-shirt can say anything about a woman's choice of partner. In a nutshell, the answer is no, not really. Among American couples, there's a slight trend for partners to be less similar in their MHCs than you'd expect by chance. Although this supports the original study, there's a suggestion that this is at least as much to do with people avoiding mates whose MHCs closely match their own as the idea that they're strongly attracted to those with very different MHC to themselves. To be fair to Wedekind, his original intention was simply to determine whether humans can detect and respond to differences in smell based on the genetics of potential mates. Like lots of other animals, we can, but it seems to play only a weak role in shaping our choices. That's not to say that smell isn't important in this context, however. With the help of a colleague, Stella Encel, I ran a straw poll asking heterosexual women how influential the natural scent of a possible partner was in shaping their perception of his attractiveness. The answers were almost unanimous: smell is a deal breaker if it's bad and a significant draw if it's right. Their views correspond with the results of surveys that indicate that a person's aroma is a critical factor in human mate choice. Women tend to rate it as being more decisive than men, but both accord it high importance. That's all the more interesting when you think of societal attitudes to smell and especially the mix of things that goes into making our smell.

The onset of puberty can be a challenging time for anyone. Alongside my friends, I underwent all manner of troubling transformations, not least in the way that I smelled. The main culprits aren't the regular sweat glands* that are distributed all over the body, but a specific type that are associated with hair follicles. The apocrine glands are concentrated in hairy parts of our bodies, particularly armpits and groins, and ooze a milky liquid that mixes into an emulsion with other substances from our sebaceous glands. While it's pretty much odourless when it first emerges, it doesn't remain inoffensive for long once it falls into the clutches of resident microbes. Among a cast of hundreds

---

* Also known as eccrine glands

of different bacterial species that call our skin home, we have two or three to thank for the rank smell that they bequeath us. One, a species of *Corynebacterium*, is a close relative of the agent that causes diphtheria. It chops and edits the complex molecules that we exude into simpler ones like butyric acid, which is sometimes described as having a smell of rancid cheese or vomit. Not to be outdone, bacteria from the genus *Staphylococcus* reconfigure our unimpeachable discharges into thioalcohols, organosulphurs with meaty or onion-y aromas with a taint of rotten eggs to them. Still more bacteria turn amino acids into propionic acid, which smells like an ill-favoured cousin of vinegar. Each of us has a slightly different mix of these whiffy ingredients, depending on our bacterial flora and hygiene, but it's generally held that men usually have an abundance of *Corynebacterium*, so have a cheesier fragrance, while women's armpits seem to favour *Staphylococcus*, lending the hint of onion to their bouquet.

None of these aromas would normally be described as pleasant; catching a waft of BO emerging from our clothes, we might register at the very least a certain awkwardness, but it wasn't always like this. Napoleon is said to have entreated his lover, Josephine, to cultivate her scent. 'Ne te lave pas, j'accours et dans huit jours je suis là,' he supposedly wrote, which roughly translates as 'Don't wash yourself, I'm hurrying back, and I'll be there in eight days.' After fermenting for over a week, Josephine would doubtless have smelled literally stunning, yet according to the story, Napoleon couldn't get enough of it. Much as we might mock this perspective, it's our modern selves who are the outliers – our evolutionary history is a story of sex and scents. For instance, have you ever wondered why we retain luxurious hair at specific parts of our bodies? One suggestion is that it's to do with the combination of our upright posture and the importance of smell for advertising ourselves. Armpits in particular are positioned near to the noses of potential smellers and are thought to be the most important contributors to our odour. The hair we grow there provides a delightfully hospitable environment for the microbes to do their stuff, while it also helps to wick moisture and volatile chemicals, announcing our odour to a waiting audience.

The fact that we start to express smells as we reach sexual maturity gives a hint as to the underlying reasons for these emanations. Our personal odours advertise our sexuality. For instance, single men smell more strongly than those in relationships, due to higher levels of testosterone. Heterosexual women judge men on their smell, and also report feeling relaxed when familiar male odours are present. Their ability to detect certain chemicals that men give off is pronounced; they can smell some components of manly aromas at concentrations hundreds of times lower than men can detect them, and this capability is at its most sensitive when they are at peak fertility. Heterosexual men, meanwhile, describe women as smelling particularly attractive at this same time; thus a chemical conspiracy draws couples together at the critical moment for conception. The scent of a woman even primes straight men for action, so that when they detect the odours it activates parts of their brain that regulate sexual responses. If the various aromas do manage to work their magic and the entranced pair surrender to the pleasures of lovemaking, the sperm home in on the egg using – what else? – their own sense of smell.

But while smell might oil the olfactory workings of attraction, it's not all about sex. Our chemical signature provides a rich source of information for anyone who smells us, and an important part of it is due to what we eat. We harness mints to freshen our breath, while treading cautiously around beans for their tendency to cause what scientists coyly refer to as flatus. Asparagus famously affects the smell of urine; eating it is a fun way of finding out which version of the olfactory receptor gene OR2M7 you have. However, unlike most other vertebrates who use pee extensively for communication, it isn't an especially important part of our smell repertoire; we're sweat specialists, or armpit aficionados, if you prefer, and diet is directly related to the chemistry of our sweat.

The good news for vegetarians is that a diet high in red meat increases both the intensity of a person's smell and the character of that odour. It's not that we sweat a meaty smell, but there's growing evidence that what we eat plays a role in determining which species of bacteria dominate the assemblage on our skin. Eat a diet dominated by red meat and you'll encourage *Corynebacteria* to make a strong contribution to your smell. Alongside red meat, processed carbohydrates also modify our scent in a way that's generally perceived as unpleasant. Given their frequent pairing on fast food menus, the red meat and carb combo is a recipe for body odour disaster.

If we want to improve the way we smell, the best way is to eat a diet rich in fruits and vegetables. These not only encourage microbes at the milder end of the smell spectrum but can even result in us having a scent that is attractive to others. Other foodstuffs that proclaim our dietary habits include herbs and spices, such as cumin, fenugreek and the dreaded garlic. I say dreaded as something of a hangover from previous generations of people from my background for whom garlic was something to be deeply suspicious of. The fact that it taints the breath was, for them, proof of garlic's subversiveness. Happily, Britons are largely free of the shackles of bland cooking nowadays, and garlic has been embraced by modern cooks as the boon that it is. While it has a certain impact on our breath, rigorous tests have revealed that eating it makes us smell more, not less, pleasant and also reduces the intensity of our scent. This might be due to the characteristics that garlic shares with other healthy foods and to the antimicrobial effect it has on our colonies of armpit bacteria.

A good diet is critical to our health, which translates into how we smell. In our olfactory perception, healthy people smell better. Changes in our bodily biochemistry caused by the onset of disease or other disorders also manifest themselves in our smell. The Greek physician, Hippocrates, first documented the value of smell in diagnostics, and it's still important today. One of the most famous examples in recent times was that of a Scottish nurse, Joy Milne, who noticed a change in her husband's smell from his usual aroma to what she described as a 'nasty, yeasty' scent. More than ten years later, Les was diagnosed with Parkinson's disease. Detecting the early signs of such debilitating conditions offers a critical advantage to doctors who can offer treatment; by simply using her nose, Joy detected the changes long before other symptoms presented themselves. It soon became apparent that she could detect the signs of Parkinson's in others. Sadly, Les lost his life to the disease, but before he passed away, he and Joy hatched a plan to put her gift to practical use. Her capabilities were initially treated with scepticism by some in the medical research community, but once she had passed their tests, they began to listen to her. As a result, a number of compounds responsible for the characteristic smell have been identified and it's hoped that a diagnostic test will soon be developed. The same potentially applies to a long list of other diseases whose presence is revealed in our body chemistry, including cancer, scarlet fever, tuberculosis, yellow fever, diabetes and Alzheimer's, among others. This simple fact offers the tantalising prospect of a new era in medical diagnostics, using sensors to make a rapid, non-invasive assessment of our health, based on how we smell.

*

While we've long suspected that our health and lifestyle affect our smell, it's perhaps more surprising to learn that temperament and character traits might do so too. Psychologists sometimes refer to the Big Five dimensions of personality, the key traits that define our character: openness, conscientiousness, extraversion, agreeableness and neuroticism. Each is influenced by hormones and neurotransmitters in our blood. For instance, serotonin and testosterone both influence aggression and dominance, while dopamine is linked to impulsivity and extraversion. This raises the question: if our biochemistry is linked to our personality, can we smell people's personality types? This was the question that Agnieszka Sorokowska from the University of Wroclaw in Poland set out to answer using the now familiar T-shirt test. A group of donors each sat a psychometric test to assess their personalities. After wearing their T-shirt for three days, they surrendered the garments to allow them to be smelled under test conditions. The results were extraordinary. Although the sniffers couldn't work out how agreeable, conscientious or open the T-shirt donors were, they were remarkably good at identifying where they sat on three scales: extraversion–introversion, neurotic–emotionally stable, and social dominance–submissiveness. The performance of the raters in this test were in some cases superior to tests that judged personality based on watching short video clips, which is extraordinary given how much importance we accord to visual cues, and how little we consciously register olfactory information.

As well our personality affecting the way that we smell, changes in our moods and emotions are mirrored in our chemical signature. It's not only dogs that can detect when people are scared; we can, too. For instance, people can reliably differentiate the smell of sweat on the clothes of gym-goers from that of novice skydivers. Moreover, when we're scared, we detect the odours of other scared people more easily – the fear spreads contagiously, like an early warning system. Beyond fear, we can also identify happy people – in this case, those who'd watched a comedy. We don't only detect the affective states of others though, we respond to them. Emotional tears contain different chemicals to those we produce to lubricate the eyes, or when we chop onions. When volunteers were asked to sniff vials of emotional tears, they showed measurable physiological responses. In essence, their bodies dialled down any aggressive urges and prepped them for empathy. We may not be aware of it, but in the background to our usual social interactions is a more subtle, yet important dialogue in the language of smell.

We're not only interested in how others smell, though. Our own aroma is a crucial part of the sensory package that we present to the world; when we smell good, we feel good, and this influences our behaviour. We can see this effect in an experiment where men were given a body spray that contained either a mix of fragrance and antibacterial agents, or a control spray that lacked these ingredients. Over several days, those using the perfumed product behaved with increasing assurance relative to the control group, who began to get a little whiffy. Female viewers asked to watch video recordings of the men judged those in the more fragrant treatment to be more attractive, even though they couldn't smell them. All of this goes to show how important our own aroma is in shaping our self-worth and confidence, and we check it more than you might think.

*

When we do sample our own smell, most of us tend to do so furtively, or in private. Being caught in the act can be embarrassing, and few people have been more embarrassed than poor Joachim Löw. He might have coached Germany to footballing glory at the World Cup in 2014, yet Löw's best known for being a bit of a Schnüffler. Standing pitchside as his team played Ukraine in the European Championships two years later, Joachim risked a swift fiddle with his crotch before having a none-too-subtle finger sniff. Alas for him, he did so in full view of millions of TV viewers. In the subsequent press conference, he confronted his shame: 'Sometimes you do things subconsciously,' he said. 'It was adrenaline and concentration.' His admission did little to contain the inevitable mockery and the media started referring to him as the 'scratch and sniff coach'. The adrenaline and concentration were at it again later in the tournament when Löw repeated his faux pas during his team's game against Slovakia. Though he was lampooned, his behaviour wasn't especially unusual, Löw's misfortune was simply to be caught in the act. The bigger picture is that his sneaky snuffle reveals something about us as a species.

Noam Sobel researches the human olfactory system at the Weizmann Institute in Israel. In 2020, he and his team produced some startling statistics about what we, and our closest primate cousins, get up to when we think no one's looking. The great apes – chimpanzees, orangutans and gorillas – touch their faces with surprising regularity, around once per minute. Though people don't do this quite so much – it's around once every couple of minutes for us – we spend almost a quarter of our waking hours with at least one hand near our noses. We're often not even consciously aware of it, as Joachim Löw suggested in his mea culpa. The question is: why? Sobel's idea was that it might facilitate a kind of olfactory sampling of ourselves and our environment. If this was so, he reasoned, we'd inhale more when our hands were near our noses. Sure enough, when he and his team measured the rate of sniffing, this was exactly what was going on. What are people sensing when they smell their hands? Partly we're smelling ourselves, both to check that we don't smell bad and perhaps to ground ourselves in our unique chemical signature. The latter reason is thought to explain why we lift our hands to our faces – a behaviour we engage in particularly when we're shocked or stressed – our own smell is reassuring to us, even though we may not realise we're doing it. It's also partly about smelling what our hands have touched, especially when we've touched someone else. For instance, Sobel noted that participants in the study smelled their hands soon after they'd shaken hands, checking out the other person's scent.

There are all kinds of reasons to check out other people's aroma; in addition to these, smell has long been employed as means of separating 'same' from 'other'. As Kālidāsa, often considered the greatest Indian writer, wrote: 'Every man has confidence in those of the same smell.' The problem comes when differentiation becomes discrimination, and when smell is used to denigrate. For centuries, the myth of the fetor Judaicus, the Jewish stink, was promulgated widely as a means to present Jewish people as being associated with Satan and depravity. William Kidd's London Directory, published in 1836, cautioned that 'their skin is so impregnated with filth as to defy the power of soap'. For those who subscribed to such views, the fetor wasn't simply about cleanliness, but it was innate, revealing something of the deep-seated corruption of Jewish people. Hitler, unsurprisingly, was a firm believer, ranting that 'Jews have a different smell' and 'the odour deterred Gentiles from marrying Jews'. The German-Jewish journalist, Bella Fromm, bravely lampooned this belief in front of senior Nazis at a reception in 1933. When Hitler unexpectedly arrived, he made his entrance kissing the hands of the ladies present, including Fromm's. 'Your Führer must have a cold,' she said acidly. 'He's supposed to be able to smell a Jew ten miles away, isn't he? Apparently, his sense of smell isn't working tonight.'

Along with Jewish people, black people have long been singled out for such hateful treatment. Dreaming up evidence of black inferiority is how slavery was justified, and smell was integral to this. Adjectives such as rank, animalistic and unclean betray an olfactory racism that became so ingrained that it formed the central pillar of a famous court case in the US, *Plessy v. Ferguson*. In 1890, the state of Louisiana instituted racial segregation on its public transport, setting aside different carriages for whites and blacks. Soon afterwards, in defiance of this, a young African-American man, Homer Plessy, sat down in a whites-only carriage. His skin tone meant that he passed as white, but by dint of his ancestry he was legally considered black. To set his challenge in motion, he had to explain to the conductor that he was, in fact, black. The nub of Plessy's defence was that you can't enforce the nonsense of segregation when you can't distinguish between black and white people. The prosecuting attorney for the State of Louisiana, John Ferguson, responded by drawing on a racist stereotype and claiming that it didn't matter whether race could be seen because Plessy's blackness could be smelled. The court found in favour of the prosecution, regardless of what to modern eyes is a ludicrous, inaccurate and blatantly racist case.

The idea that races smell different from each other is little more than a myth – after all, the differences in the genotypes of any two people are vanishingly small. There is, however, one caveat to this. The ABCC11 gene is heavily involved in both body odour and, oddly, earwax. Around 40,000 years ago, a new variant of this gene appeared in East Asia. It has the effect of toning down both the activity of apocrine glands and their exudations of the molecules that armpit bacteria turn into BO, while also drying up earwax. Since it first appeared, the variant has become firmly established in Japan, Korea and China, where between 80 to 95 per cent of people carry it, compared to less than 3 per cent in populations of European or African descent. It's difficult to know for certain why this gene might have gained such a strong foothold in Asia, but the working hypothesis is that as people moved from Africa into colder climes, a gene that restricted sweating could have survival benefits. Whatever the reason, the effect is to reduce the armpit odours of East Asians, though a breakdown of the cocktail of chemicals we produce revealed no significant differences between ethnicities.

Any other differences between us come not from genetics, but from differences in diet and culture. For example, legend has it that when European traders first reached Japan, the locals were appalled at the smell of the foreigners. Differences in diet, and particularly the Western fondness for dairy products, alien to the Japanese at the time, probably contributed to this. Borrowing the English term for what they thought to be the chief culprit for the Westerners' unconscionable stench, the Japanese referred to the Europeans as *bata-kusai*, literally 'butter-stinkers'.

Aside from the change to that one gene, there are few differences between us. So why did such powerful and obnoxious stereotypes about smell emerge and why do they persist? An interesting insight into the psychology of this was provided by a team of researchers from the Universities of St Andrews and Sussex, led by Stephen Reicher. In their study, students were given the ubiquitous T-shirts to smell and asked to report their level of disgust as well as measuring how quickly they rushed off to wash their hands afterwards. The twist was that some of the T-shirts were emblazoned with the logo of their own university, while others carried that of a rival institution or no logo at all. Smelling a stranger's armpit fug isn't the most pleasant experience and accordingly, the students were somewhat revolted by it. The difference came when they believed they were sniffing a T-shirt from their own university compared to another; they found their own ingroup far less disgusting. We seem to find the emanations of people we're close to, or have more in common with, far less repugnant than we do those of outsiders or strangers.

Armpit aromas might be bad, but they're not the worst odours we produce; each of us farts on average ten times a day. To let off in private is one thing, to do so in company is considered poor form in almost every culture – there's nothing like cutting the cheese during polite conversation to cause one's social stock to plummet, especially if there isn't a nearby dog to cop the blame. Although we all do it, other people's farts are much more appalling than the ones that we produce. A small part of this might be to do with the fact that our own emanations don't take us by surprise, but mostly it's a deep-seated response to the threat of contamination and disease that people, particularly strangers, present. After all, we don't often make ourselves ill, but we can be infected by others. Consequently, we're wired to avoid anything that jeopardises us; even though farts are extremely unlikely to transmit illness, their smell is redolent of faeces, which really do pose a threat.

*

Disgust is an evolutionary essential. Without it, we'd risk contaminating ourselves with all manner of noxious chemicals, conditions and critters. When do we most consciously use our noses? Probably when we retrieve a murky-looking container from the furthest recesses of the fridge and find that it's gone beyond the best before date. The brave among us might try a sniff to judge whether it's worth a risk. If it isn't, the smell will let us know because our olfactory system is primed to steer us away from danger. Bad smells trigger the brain far more than pleasant ones.

The link between emotions and smell goes both ways. If you're expecting the jar that you've recovered from the fridge to smell horrific, your anxiety will play a part in making this so. Consequently, a perfectly innocent foodstuff might be discarded on the basis of a jaundiced perspective. In an experiment to test this, the researchers Rachel Herz and Julia von Clef took a range of substances and asked people to rate how they smelled. The trick was that each substance was given a positive label and a negative one; people were asked to smell both, with a week in between the two smells, without knowing that they were smelling the same thing. For instance, pine oil was labelled either as 'Christmas Tree' or 'Toilet Cleaner', and a blend of isovaleric acid and butyric acid was designated either as 'Parmesan' or 'Vomit'. As you might expect, people rated smells as more pleasant when they had been primed by a positive label, but the considerable differences in their ratings of exactly the same smell demonstrate that we are highly prone to prejudice.

The nose can be a hostage to our expectations, sometimes to the extent of olfactory hallucinations, something that was demonstrated to dazzling effect by two daring TV hoaxes. In 1965, the BBC aired an interview with a supposedly eminent professor who claimed to have perfected what he called smell-o-vision. He added onions to his miraculous machine, telling viewers that molecules of smell would reach them best if they stood 6 feet from their set. Despite it being April Fool's Day, viewers wrote in to complain that not only had they smelled the onion, but that it had caused their eyes to stream. A few years later, a lecturer at the University of Bristol, Michael O'Mahony, drew the attention of television viewers to a large cone in the studio, rigged up with various wires, and told them that he could transmit an odour in sound form using vibrations, so that when they heard it, they'd smell a 'pleasant, country smell'. Rather than elaborate further, he asked the viewers to contact the TV station to describe what they'd smelled. Dozens of people wrote in, over half of whom thought they'd smelled hay or grass. Some even complained furiously that the experiment had given them hay fever.

Smells influence everything from our strategising to our social behaviour. A test conducted by researchers from the University of Liverpool and Unilever in 2015 examined the propensity of people to gamble while being wafted with various aromas. Smells of either jasmine or methylmercaptan, a key ingredient in bad breath and farts, were floated among the unwary subjects to see how it affected their choices. As the experimenters predicted, the stench of mercaptan made people more cautious, cutting their losses and avoiding risk. Smells can be portentous signals of a change in the status quo, and while a small financial loss isn't comparable to our ancestors' experience of confronting a sabre-toothed tiger, the scent of something aversive can activate our deeply ingrained instinct to withdraw from danger.

Odour can also influence our broader moral judgements. Yoel Inbar, a researcher from Cornell University, famously tested the feelings of young, heterosexual subjects towards homosexuals and the elderly. The twist was that the participants were asked to complete their questionnaire either in a standard room, or in a room that had been doctored by the judicious application of fart spray. Although the sulphurous odour was unremarked by the experimenters, the results were unequivocal – the attitudes of those in the stinky room became more prejudicial. In other words, our senses profoundly shape the way that we think and interact with the world.

You might be starting to wonder whether psychological research into odours involves anything other than stewing people in bum gas to see how they react. Researchers do sometimes turn their attention toward the pleasant end of the aroma spectrum; when they do, they make their subjects happy. For instance, most people would agree that citrus is a pleasant smell. It's zesty, appealing and fresh, a virtue that has led to manufacturers using it to promote their products. Lemons in particular have long been used for cleaning, and now you can revel in the smell of these fruits as you wipe down a grimy kitchen work surface or clean a shower. In 2012, two Dutch researchers, René Wijk and Suzet Zijlstra, measured not only the opinions of volunteers exposed to citrus aromas but a battery of psychological, physiological and behavioural responses. As well as enjoying the smell of citrus, the volunteers became more active, performed better at a simple test, and expressed more positive emotions. They also made different choices at a buffet, choosing to eat more mandarins and less cheese.

While advertisers have traditionally appealed to us via our sight and hearing, cunning marketeers are leading a growing movement to incorporate the other senses, all the better to part us from our cash. They're able to do this because our senses dictate our behaviour. For instance, the addition of a subtle citrus fragrance to an auction room increased the prices that people were prepared to pay by over a third, and the addition of a pleasing odorant to a casino resulted in punters staying for longer and spending 50 per cent more on slot machines. The scope of the change that can be induced by a simple exposure to a smell demonstrates the power of aromas to shape our perceptions, yet often in a way that occurs below our conscious appreciation.

Conscious or not, smell shows promise as a means of regulating pain. The mere fact that odours can be perceived quickly and without requiring a great deal of attention makes them strong candidates for use in alleviating pain. Pleasant smells both improve the mood and reduce the anxiety of people undergoing minor medical procedures. The problem is that unpleasant smells do the opposite, and their effects are much more powerful. Many of us are familiar with the smell of hospitals. It's not necessarily objectively awful, though we might learn to associate it with fear and stress, which in turn can increase the impact and intensity of pain, creating a kind of vicious cycle. Learning new smells and then associating them with either positive or negative experiences is something that we excel at. Rachel Herz of Brown University in the US has conducted studies which reveal how exposing people to novel aromas while they're engaged in a challenging or frustrating task leads them to link that smell with their annoyance, reducing their willingness to attempt a different task the next time they smell it.

As a quick aside, the feelings of relaxation that we experience when smelling pleasant odours form the basis of aromatherapy, which although invented in its present form in the 1920s, has been practised in one form or another since ancient times. However, aromatherapy differs from its scientific equivalent, aromachnology, in that the latter relies on a rigorous, objective approach to the study of how smells affect us. In short, aromachnology doesn't set out to debunk aromatherapy entirely, rather it seeks to constrain excessive claims and to move from anecdote to data. Although there is evidence that aromatic compounds can get into the bloodstream through the mucosal lining of the nose and lungs, they tend to do so only at very low levels, and certainly nothing like the dosages we're used to with more traditional pharmaceuticals. Moreover, even if they do get into the blood, we'd expect it to take a quarter of an hour or more to travel across the blood-brain barrier to have a direct effect on the mind. In contrast, most people report that odours act much more quickly than this, often within a few seconds. Overall, this suggests that people's responses to smells are primarily emotional and psychological. This doesn't negate the value of smelling lavender oil, or whichever product you prefer, it's just to say that the means by which it works isn't quite the same as medicine.

What works best as a mood-improving smell depends on who's smelling it. The differences between individuals can be profound. For instance, the Daasanach people of East Africa are pastoralists, whose society is founded on cattle. For them, bovine is the finest scent; accordingly, they anoint themselves with manure, urine and butter. While I wouldn't necessarily share the wholehearted commitment of the Daasanach, I agree that there's something pleasant about the warm, sweet scent of cows. Among Westerners, there's a cultural schism between Europeans and Americans in relation to the smell of wintergreen, an evergreen shrub native to North America. It has a kind of minty, antiseptic aroma and is used to flavour gum, candy and toothpaste in the US, as well as a drink of inexplicable popularity, root beer. Americans, by and large, have only positive things to say about the smell. Meanwhile, a European encouraged to take a sip of root beer will tend to recoil in horror. How could the same smell provoke such different reactions in similar cultures? It might be to do with the associations that the smell conjures. To me, and to many Europeans, wintergreen smells like the kind of mouthwash you only experience at the dentist. To Americans, it's a familiar smell from childhood, one that's largely associated with treats and positive connotations. Yet for all the discrepancies between our preferences, focussing entirely on these would obscure the broader point that we're more likely to agree than to disagree on the question of whether something smells good. If you strip olfaction down to its biological basics, you might conclude that whether something smells good or bad is decided by its potential to help or harm us. But smell is more complex than that, and layered on top of the biology are a host of emotions and associations that give each aroma a meaning.

*

Perhaps more than any other sense, olfaction allows us to reach back into our past and to recover autobiographical memories. Times and places, even emotions, are anchored in our subconscious by particular odours and can be brought into sharp relief by even the faintest whiff. This phenomenon is often labelled the Proust Effect, in honour of Marcel Proust's description of *déjà vu* upon tasting a tea-soaked cake in *Remembrance of Things Past*. Smell, as well as taste, seem to trigger these vivid recollections to a greater extent than other senses; it's been suggested that this might be to do with the way our brain is organised. In particular, the neural pathways of the olfactory sense link closely not only to the olfactory bulb, but also to the limbic system, which is intrinsic to memory formation and recall.

This intimate connection between aromas and reminiscence seems to be a common experience. Though the specific odour varies, almost all of us experience that strange sense of being relocated back in time upon catching an old and familiar scent. For instance, I have only to smell a hint of aniseed and I'm transported back to childhood, sitting by my mum's side in our old car as she drives us down the high street, eating one of the sweets she used to treat herself to every now and again. As with my own example, for most people the experience provokes memories of a bygone stage of their lives. Additionally, these memories seem to be more vivid and emotionally charged than those based on visual cues. According to Simon Chu and John Downes of the University of Liverpool, odours seem in most people to recall memories that were established when they were aged between six and ten. This aspect in particular is fascinating because it contrasts with a pattern known at the reminiscence bump. When people above the age of forty are asked to describe their memories, the richest recollections tend to be drawn from between their early adolescence and mid-twenties, with a peak in their mid- to late teens. The reasons for this are manifold, involving the development of personal identity, peaks in cognitive ability, and exposure to new and exciting things. Odour memories, however, are established well before this, which might imply that there's a deeper significance to their establishment.

Recent research suggests that the sense of smell is intrinsically linked to navigating one's environment. Comparisons among closely related species reveals a link between brain morphology and the distances over which they tend to roam. In particular, the more widely a species travels, the larger its olfactory bulb and hippocampus. This latter structure is a part of the brain that plays a fundamental role in learning and memory. It's a constituent of the limbic system, which, as we've already seen, is intimately associated with olfaction. Navigating by smell isn't solely about homing in on a particular odour, as with salmon or pigeons; it's about creating a kind of olfactory map of an area in the brain, a smellscape that guides an animal in its wanderings. This ability has long been used by sailors and early aviators to navigate over long distances, for instance using chemicals such as terpenes that are given off by forests, or dimethyl sulphide released by marine algae. A recent study by Lucia Jacobs and her co-workers at the University of California, Berkeley, demonstrated that blindfolded people can accurately orient themselves and find their way around a room using a mosaic of different smells.

In 2014, the neuroscientists John O'Keefe, May-Britt Moser and Edvard Moser were awarded the Nobel Prize for their work in understanding how the brain encodes spatial information. Building up a map of the environment involves specialised neurones in the hippocampus, known as place cells, working in conjunction with so-called grid cells. As an animal moves around in its environment, different place cells activate. Alongside this, grid cells map out the coordinates of that environment, and represent its key elements. Working in collaboration, these two cell types provide a kind of built-in GPS. While many sensory cues provide the input needed to develop and maintain this neural GPS, olfaction plays a critical role. As well as the grid and place cells, an adjacent structure in the brain, the anterior olfactory nucleus, communicates with both the olfactory bulb and the hippocampus. This combination contextualises odour information to create memories concerning when and where they were encountered. Ultimately, these odour memories and the feelings they evoke seem to be related to our need to learn about our environment at an early age.

As well as helping to organise our memory, our sense of smell acts as a valuable early warning system, alerting us to the presence of threatening contaminants and letting us know when something smells 'wrong'. But although the human nose is a marvel in many respects, we've sometimes had to recruit other animals to do sensing on our behalf. Think of truffle hunters working with the aid of fungus-sensitive pigs, or the sniffer dogs used to detect all manner of contraband, as well as explosives. A scaled-down version can be seen in the WaspHound, developed by researchers at the University of Georgia. Tiny wasps can be trained to detect infinitesimal concentrations of anything from crop disease to dynamite. When they do so, they change their behaviour, becoming highly animated inside their containers. The wasps are trained by being given grains of sugar in the presence of whichever smell they're being trained to recognise. Once they've learned, it's the prospect of sugar that gets them excited upon exposure to the smell.

The exquisite sensitivities of animals can be used to protect us in other ways. The Warsaw clams, which sounds like a fabulous name for a band, are legendary molluscs in Poland. They inhabit the Fat Kathy pumping station, sitting in the water supply and passing time in the manner of clams everywhere, which is to say, sucking water in and filtering out tasty particles. Their usefulness comes from the fact that they are extremely responsive to pollutants. When they detect heavy metals or other nasties, they close up, which sets off an alert in the pumping station and thus prevents foul chemicals getting into the water supply.

As ingenious as all of this is, it's not an approach that lends itself to every application – and this is where technology comes in. While developing sensors that can perform to the level of the human olfactory system has presented enormous challenges, we've finally picked up the gauntlet laid down by Alexander Graham Bell and we're beginning to measure smells. Electronic noses have been around since the early 60s, though it's only in recent times that they've become more than a curio. At the heart of the machine is an array of chemical sensors, each of which detects a specific type of odorant molecule. The information from these sensors is then passed on to software that performs pattern recognition on the overall blend. In this way, the technology behaves similarly to the nose on your face. What it can't do, as yet, is extract meaning from an odour; this remains the preserve of our brain.

In the case of the food industry, the 'e-Nose' can determine whether something is safe to eat, or if a tell-tale signature of bacteria indicates that the food has started to spoil. They can provide quality control for everything from raw ingredients to ensuring that the finished product meets expected standards. Indeed, given the colossal amount of food waste that's generated around the world, it may soon be possible to integrate basic sensors into food packaging. These might change colour when the contents are no longer edible, ending the gastric Russian roulette that we play when deciding whether to ignore a use-by date. Something like this is already on the market for judging the state of fruit. The ripeSense alerts consumers when avocados and pears are perfectly ripe. Beyond protecting what we eat, e-Noses can be used in medicinal diagnoses, for protecting the environment and they may eventually make sniffer dogs redundant. The technology behind them may one day offer a solution that restores the sense of smell to those with anosmia.

The future isn't all about detecting smells; it's also about creating them. Work is already underway to develop bespoke perfumes based on individual preference. Key in a few details and a perfume printer will pop out a whiff of scent for your delectation. Using similar technology, it may one day be possible to realise the dream of 'smell vision'. The idea of conjoining film with appropriate aromas has a long history but has always faced significant hurdles. For one thing, it's a challenge to make the smell appear and then disappear sufficiently quickly. With wearable technology, however, it might be possible to excite our sense of smell alongside vision and hearing in a coherent way. Not that you'd want it for all films; who wants to watch a zombie movie with the smell turned up.

One new smell is what's been called 'olfactory white'. It's the scent equivalent of white noise, a combination of a broad spectrum of different smells that cancel each other out to leave a kind of non-descript scent that's neither nice nor nasty. When you amalgamate thirty or more diverse odours, the olfactory system becomes saturated, overloading the brain with information, and the ability to detect different components is lost. The idea has genuine applications. For instance, at present when we're faced with a foul smell, we try to cover it with something like air freshener. All that happens, in my experience, is that the stink and the 'fresh' scent battle against each other and everyone loses. By contrast, olfactory white could be used to completely vanquish unwanted aromas, leaving public toilets and hospitals smelling entirely neutral.

While such advancements are exciting, we shouldn't lose track of the wonder of our sense of smell. It was the first sense to develop back in the ancient history of life and it is now the most widespread among the organisms with whom we share the planet. The simplest creatures can detect chemicals in their environment; just about every bacterium can do it, as can fungi like yeast, and even plants. This foundational 'chemosense' became olfaction as creatures' brains evolved to process sensory information; ultimately, dedicated structures like the nose developed to hone and localise scents. The deep-seated idea that humans have a poor sense of smell is gradually being overturned, alongside the idea that it's of secondary importance. Smell is fundamental to our lives, used by babies from their first day of life when glands around the mother's nipple secrete chemicals that guide them to suckle at the right place. Evolution tuned our olfactory palette to odours that indicate opportunity or danger. It's not unreasonable to assume that those scents to which we are most sensitive are those that played the most important roles in human history. Accordingly, the low detection thresholds for the aromas of ripening fruit, for sulphurs that indicate fire, and geosmin* that informs us about water contamination can be seen as having been critical to our survival. The simple fact is that we are successful as a species because of our sense of smell, and it continues to protect us and give us pleasure. It's the sense that governs our appetites, marshals our sex lives and provokes our finest emotions.

---

* Literally, 'earth smell', this compound is produced by blue-green algae and other bacteria, which at high levels can potentially cause health effects in humans.

# Accounting for Taste

*The myriad of flavours explode on my tongue,*
*shimmy through my mouth, slap my taste buds*
*and call them filthy bastards, and I love it.*

*– Stacey Jay*

I'm sitting with friends in a warm bar, sheltering from a blustery winter's day in Iceland, and in a state of nervous excitement. I'm excited because I've persuaded my reluctant companions to make time for a very particular experience. The nervousness is because the experience isn't likely to be pleasant. The term glutton for punishment was never more apt; I'm here to try the Icelandic delicacy, *hákarl*, or fermented shark.

When it comes to food, Icelanders have had to be creative. For centuries, they've had to be open to options including puffins, sheep heads, sour rams' testicles* and some rather large fish. Greenland sharks, one of the North Atlantic's bounties, reach up to seven metres in length and a tonne in weight. Regrettably for the more discerning diner, while most of the animals we eat expel toxic urea from their bodies in their pee, most sharks store it in their blood. Not only does this add a certain taint to the flavour, but tucking into fresh shark is risky: too much can kill you and even a small amount can render you 'shark drunk'.

---

* No wonder the rams are sour.

147

The wily ancestors of the modern Icelanders devised a solution for this – they buried the shark in a shallow grave of pebbles. You could reasonably argue that the wisest option would be to leave it there, but Icelanders are tougher than that. After three months, hordes of bacteria have enjoyed a bonanza: there's half a trillion of them in every teaspoon of shark. Not only have these bacteria putrefied the carcass, they've also converted the urea to ammonia. At this point, the Icelanders consider the shark done to a turn, whereupon they exhume it and hang it in great festering, meaty curtains to dry. After that, all that remains is to eat it.

The *hákarl* is delivered to me in chunks, sealed in a Kilner jar lest its terrible smell frighten the horses. My friends, who stubbornly refuse to participate, watch on as I timidly unfasten the container and retrieve a gobbet with the toothpick supplied. There's no going back now. I pop the heinous morsel in my mouth. I don't gag, though many first-timers do, apparently. A wave of flavour breaks over my tongue, a gustatory collage of particularly disreputable public toilets. There's a note of elderly fish, swimming valiantly against the lavatorial flow. The texture is troubling, too, a kind of rubbery malevolence. To sum up the experience, I'd probably go with 'vulcanised litter tray'.

Human cultures all around the world have experimented with the foods that they found around them. In many cases, they may be poisonous until they've undergone some preparation. *Hákarl* is one example, but it also applies to much more familiar foods such as the world's most popular root vegetable, the potato. The wild ancestor of the familiar spud grows in the Andes and is laced with toxic compounds like solanine and tomatine. Cooking has little effect on these toxins. so how did people manage to bypass the potato's defences? It's possible that, centuries ago, the local people learned a trick from animals such as the guanaco, a close relative of the llama, which licks clay before eating poisonous plants. Essentially, the clay binds to the plant toxins, rendering them inert and allowing them to harmlessly pass through the animal's body. Early Andean potato fans likely did something similar, using a kind of muddy dipping sauce to accompany their toxic tubers. Since then, of course, selective breeding has given us an innocuous potato, one that's less likely to send you to an early grave. Nonetheless, even today, poisonous varieties of potato are still in favour with a hard core of devotees, partly because their resistance to frost means they can be grown at altitude. At the markets in Peru and Bolivia, you can buy the essential condiment, clay, alongside the deadly spuds.

Like the potato, many plants employ chemical defences to discourage browsing animals. Some, like belladonna, hemlock and oleander, are notorious for their potency. Ricin, produced by the castor oil plant, was famously used in the assassination of the Bulgarian dissident Georgi Markov by the KGB in 1978. While few plants pack this kind of punch, there are plenty that can cause severe effects. This being the case, how did our ancestors avoid being poisoned as they explored the culinary potential of unfamiliar plants? This is where taste comes in.

Anything that infiltrates the body is potentially harmful, and the mouth represents the front line of our defences. Consequently, our mouths are complex, multifaceted sensors whose function is to protect us. When we first take a bite of a new food, we evaluate its flavour, making a sensory assessment of a food's chemical composition. This assessment is passed on to the brain and processed rapidly. If the feedback is good, we keep on chewing; if not, we can quickly spit it out, ideally before we come to any harm. Taste, then, is a sense that allows us to explore and enjoy the nutrients in food, as well as keeping us safe from harm. However, while this sounds straightforward enough, what we actually think of as taste is a composite of different senses, working in collaboration.

\*

One of the most exciting things about travelling is tasting the foods of different cultures. Through them, I can gain a sensory insight into a culture. That's not to say that *hákarl*, for instance, is the distillation of everything that Iceland represents, rather it's a small, but important, facet of their history and custom. I've been fortunate enough to try all manner of things, from beaver tails* in Canada, the maize porridge known as *ugali* in Kenya, and the one I struggled with more than anything else, *nattō* in Tokyo. This latter is made of fermented soybeans, which (to me, at least) has a punchy bouquet of aged sock, but it was its stringy, slimy texture that ultimately defeated me. When we talk about taste, we often use this as a shorthand for flavour perception, which is often said to emerge through a combination of taste, smell, texture and something known as chemesthesis.

Strictly speaking, taste is the part of the sensory experience that's performed by dedicated taste cells that are dotted around the mouth. These cells come in different forms, corresponding to what we might think of as the five primary tastes: sweetness, saltiness, sourness, bitterness and umami (or 'savouriness'). While these are our perceptions, it's helpful to dig down a little and to think in terms of what exactly each taste cell is detecting. It's obvious what sets off salt receptors, but it's worth mentioning that it's mainly sodium salts that provide the salty sensation. Other metal salts, like potassium, can be useful stand-ins for the health-conscious, but they're not as intensely flavourful and can even be sensed as bitter. Regular table salt, properly known as sodium chloride, gets split by saliva into sodium ions and chloride ions. The increase in the concentration of sodium ions in the mouth when we eat salty food is what sets off the nervous impulse that the brain translates as the impression of saltiness. Despite concerns associated with an excess of salt in the modern diet, sodium ions in moderation are essential to physiology. It tastes good because it is good.

---

* This is a pastry, not part of the beaver.

Sour tastes are the perception we gain from eating or drinking things with a low pH, in other words acids. The more hydrogen ions there are in a substance, the more acidic it is, and we register sourness in a similar way to that in which we detect salt – the taste cells are responding to an increase in the concentration of ions. Although sourness can be an indicator that food has gone off, sour foods are a mainstay of the human diet, and include most fruits and many vegetables, as well as bread and rice. Even milk, cheese, eggs and meat are slightly acidic. These foods aren't equally sour though; while milk hovers just below neutral, lemons and oranges are thousands of times more acidic, as are lots of fizzy drinks. When foodstuffs have a particularly low pH, they can taste distractingly sour. A teaspoon of fresh lemon juice, for instance, is enough to make even the most stoic among us grimace. It also sets off a salivary flood within the mouth, which is essentially the body's attempt to dilute the acid. This sourness can be masked, however: add lots of sugar and the sour taste retreats. This is because we're wired by evolution to tune into sweet substances – they're energy-rich and highly rewarding, and until relatively recently in the history of our species were comparatively rare. Adding lots of sugar to a sour food relegates the acid to the background, while the sweetness takes centre stage. What it doesn't do is make the sour food any less acidic, which is why your dentist is always prattling on about the dangers of fizzy drinks, since the sugar-acid combo is a bit of a disaster for the teeth.

While salty and sour tastes are the result of taste cells registering changes in ion concentrations in the mouth, the remaining three tastes emerge through a different mechanism. These work a little like smell, in that food molecules have to actually bind to a sensory receptor to set off a nervous impulse. We're used to thinking of sugar as being synonymous with our perception of sweetness, but in fact thousands of different chemicals, including salts of metals like lead, as well as certain alcohols and even amino acids all give the perception of sweetness. Our craving for sugar, in combination with the fact that so many different materials can trigger the sweet receptors in the mouth, is the reason that food manufacturers have developed products that not only mimic sugar, but that we perceive as being much sweeter. Aspartame, sucralose and neotame are respectively 180, 600, and 8,000 times as sweet as the real thing. Just how sweet something is depends on how well it binds to the sweetness receptor. The better the receptor's fit with the molecule and the more sites at which they interact, the sweeter we taste it to be. Artificial sweeteners are tailored to fit our sweetness receptors as snugly as a hand in a glove.

Aspartame and its chemical cousins are by no means the only substances that can hoodwink our sense of taste. Perhaps the most remarkable of these is miraculin, an extract from the miracle fruit that has been cultivated for centuries in West Africa. When it is eaten alongside something sour, something weird happens: the sour food tastes sweet. It makes lemons taste like the sweetest oranges and converts salt and vinegar crisps from a savoury snack into a kind of briny pudding. In doing so, it's not transforming the composition of the food – this remains unchanged. Instead, it alters our perceptions by converting the acid taste to a sweet one.

In a similar way, the artichoke contains a chemical known as cynarin, which temporarily locks on to our sweetness receptors and seems to deactivate them. Sipping a drink after eating an artichoke washes the cynarin off the receptors and triggers the brain into thinking that you've tasted something sweet. A more familiar taste trick is performed by a common ingredient of toothpaste, sodium lauryl sulphate (SLS). This mild detergent is the reason that if you eat or drink anything in the immediate aftermath of cleaning your teeth, the flavour is so odd. Like all detergents, SLS attacks fat molecules, but when it affects the fatty membranes of our taste cells, they go a little haywire. It puts a temporary block on sweetness reception while also messing with bitterness detection. The result is that if you have a sip of orange juice soon after brushing your teeth, you taste none of the sweetness, and the sour flavour of the drink comes across as bitter. Keep drinking and you'll wash the SLS away and normal service will be resumed.

Until fairly recently, it was common for textbooks to list just four main types of primary tastes, missing out on the fifth altogether. Over a century ago, Kikunae Ikeda, a chemist at the Tokyo Imperial University, coined the word 'umami', which roughly translated means 'essence of deliciousness'. Despite its long history, it wasn't until 2002 that dedicated receptors were identified for umami, gaining it the seal of approval from the scientific community as the fifth taste. Regardless of this, umami has long been a part of culinary cultures. The ancient Romans used fermented fish sauces to add piquancy to their meals, while the Byzantines and Arabs achieved the same effect with murri, a condiment made from barley, and soy sauce has been used in China for almost 2,000 years. Even with this rich history, umami remains the hardest taste to define. The best I can muster is to say that it is savoury, intense and delicious. It's found in a whole range of foods, including meat and fish, cheese, mushrooms and tomatoes, and the kelp seaweed, known as kombu in Japan, which was the basis of Ikeda's early studies. In all these cases, our taste receptors are picking up on amino acids, such as glutamate. Amino acids are the fundamental building blocks of proteins and an essential part of our diet; our predilection for umami is the body's way of steering us toward a critical food source.

Despite the savoury hit that it adds to our dishes, umami is regarded with suspicion by some. Building on Kikunae Ikeda's discovery that glutamate salts carried the savoury flavour, production began shortly afterwards of what is today perhaps the most vilified food additive in the world: monosodium glutamate. MSG, as it came to be known, was a huge commercial success as a food additive across the globe, yet it retained an association with its origins in the East. It had long been established and widely accepted when, in 1968, everything changed: a letter was published in the *New England Journal of Medicine* claiming that MSG was the cause of heart palpitations, headaches and chest pains. The claim triggered a wave of anecdotes, gaining both widespread attention as well as a name, 'Chinese Restaurant Syndrome', that stigmatised both MSG and Asian cuisine. Early studies appeared to back the claims up, and MSG underwent a transformation from panacea to poison. What those early studies lacked, however, was scientific rigour. Most important of all, the subjects knew when their food contained MSG, and when it did the researchers had added disproportionately enormous quantities of it. Given the hysteria surrounding MSG, it's little wonder that many of them developed the symptoms. This is a well-documented phenomenon, known as the 'nocebo' effect, the polar opposite of the better-known placebo effect, and occurs when people develop symptoms even in the absence of any actual cause, based on their negative expectations. Research carried out using more exacting protocols, including blind tasting, produced very different findings. Even among people with a self-reported sensitivity to MSG, the incidence of any ill effects was vanishingly small. During the last fifty years, when MSG has been demonised, people have been merrily consuming glutamate as part of their normal diet, since it occurs naturally in so many foods, yet the additive remains under suspicion. Even today, many restaurants, especially Asian restaurants, advertise 'no MSG' in their offerings, which only serves to support preconceptions. Of course, people are free to decide what they eat, but such attitudes toward MSG are more about belief than science. Its poor reputation in some quarters goes to prove the old maxim: people prefer the bunk to the debunk.

While opinions, if not data, vary on the question of whether MSG is toxic, our sense of taste is geared toward detecting other substances that are. We don't often think of bitter tastes as being especially important, yet the mouth is a bitter specialist. For instances, there are just three receptors dedicated to the detection of sweet and umami tastes, while at least twenty-five contribute to the perception of bitterness. At the root of this is the fact that eating has long been a dangerous occupation. Bitter tastes are especially important for their association with plant toxins. Plants are loath to be eaten, but their capacity to run away is limited. Consequently, they marshal their defences in the form of chemicals that pack a punch and these are almost exclusively bitter tasting. The exception to this is when plants are ready to be eaten, such as when they need their seeds to be distributed. At this point, they withdraw the protective bitter chemicals and add sugar to encourage animals to do their bidding.

Not all bitter plant compounds are injurious to us. Some, such as those in coffee, cocoa and the hops that we add to beer are appreciably pleasant. Then again, bitter and dangerous chemicals reside even in some of the most innocuous foodstuffs. Cassava contains cyanide and soybeans incorporate a substance known as saponin that destroys red blood cells. Even the humble turnip contains a chemical that suppresses hormones within the body. It's important to point out that you'd need to eat vast quantities of any of these to be on the danger list. The same isn't true of plants such as belladonna and hemlock that can kill rapidly at small doses; happily, our sense of taste arms us with the ability to detect dangerous chemicals. Indeed, our sensitivity to bitterness is over 1,000 times more acute than to sweet or salty flavours, alerting us to the presence of even tiny quantities of poisons in our mouths.

Bitter is the counterpart to sour, and while sourness comes from acids, bitter tastes are a signal of alkalinity. Foods with extreme pH very often cause us to reject them, but this is especially the case with bitter foods. We're born with an aversion to bitterness, to the extent that rejection of foods that taste like this is close to a reflex response. The problem with this is that many of the compounds found in vegetables, as well as in medicines based on plant chemicals, have a bitter component that children in particular loathe. The bitterness of Brussels sprouts and other cruciferous plants comes in large part from glucosinolate, a substance that many kids and some adults find revolting. This has led some food manufacturers to investigate the possibility of masking the bitter flavour. One approach is to add sweetness; this works to some extent, but our sensitivity to bitterness means that quite a lot of sugar is needed to achieve the desired effect. As an alternative, salt has some valuable effects. It's often said that salt enhances flavour, but it would be more accurate to say that it selectively suppresses particular tastes. Salt seems to be especially useful in toning down bitter notes in food, giving rise to a perception that other flavours are enhanced because of the way that they emerge from bitterness's shadow in the presence of sodium chloride. Of course it brings with it a new set of concerns, so one solution potentially begets another.

The combination of using sugar to balance acidity, salt to restrain bitterness, and umami to bring out savouriness has been a mainstay of cookery for centuries. An idea often expressed is that, by paying attention to the interaction between these five tastes, we can create dishes that ascend toward perfection. Among scientists, too, it was thought that the inclusion of umami represented the final piece of the jigsaw. Recent research, however, suggests that hidden among the mouth's better-known receptor cells are many more whose functions have long been shrouded in mystery. For many years it has been argued that since we have specific receptors dedicated to two of the three main macronutrients, carbohydrates and proteins, it would be logical for us to be able to detect the third: fat. Though the debate continues, a receptor has now been identified. G protein-coupled receptor 120 – GPR120 to its friends – is activated by the presence of fatty acids, the building blocks of fats. And, just as with our other primary tastes, when a fat is placed on the tongue, around a tenth of a second later there's a measurable response in the brain. The case for including fat as a primary taste now seems compelling.

Other enigmatic taste receptors in the mouth are also starting to give up their secrets. Some claim that the metallic tastes that manifest when we get a bang on the head, and the starchy flavours that we perceive when we eat pasta or rice, are distinct tastes in their own right. It's also been argued that that the elixir of life itself, water, might have its own specialist taste receptors. Insects, for instance, seem to possess something along these lines and evidence is accumulating that we have dedicated water sensors throughout the body that have much in common with taste receptors. Water might not taste of very much, but if you've ever tasted deionised water – in other words water with everything but the $H_2o$ removed – it tastes different, some say slightly bitter. Much of water's flavour comes from the tiny amounts of minerals that it contains, and in particular metal ions. Our salt receptors detect some of these, but in addition evidence is emerging that our tongues are armed with the specific ability to sense the presence of calcium, another substance crucial to the body's function. The surprise, however, is that we don't necessarily taste calcium in the foods we most associate with it; fats and proteins in milk and cheese bind to the calcium and stop us from detecting it. We're more likely to taste it in green vegetables, such as kale and other members of the cabbage family, which contain this mineral in high concentrations.

Where calcium receptors really seem to come into their own is in the detection of something known as kokumi. For anyone still struggling to come to terms with the idea of umami, kokumi is going to come as a body blow. Like umami, it was first identified in Japan. Its name, 'kokumi', translates as 'rich taste', which is somewhat ironic because kokumi doesn't really taste of anything – it's more a feeling, sometimes described as 'mouthfulness'. It gives more familiar tastes a boost, drawing out their flavour and providing a roundness to the experience. Despite the fact that calcium receptors seem to play a part in its detection, the molecules that yield the kokumi sensation are a range of different peptides – sort of apprentice proteins – that emerge from slow-roasted meats, aged cheeses and fermented foods. Describing kokumi is like trying to nail a jelly to the wall, but it's the guiding principle behind long established practices such as hanging meats and maturing cheeses. We know they taste better for it, and kokumi is a key reason why.

So will we add fat, starchy, metallic, water, calcium and kokumi to the five primaries? It's too early to tell – for all that we might think taste is settled and understood, the truth is that we have much to learn.

*

Within the mouth, clusters of 50 to 150 taste cells are packed together within tiny, raised bumps: the taste buds. Most people have somewhere in the region of 5,000-10,000 taste buds, the vast majority of which can be found on the tongue, with the remainder on the inside of the cheeks, the roof of the mouth and the throat. Each of them has a miniscule pore facing out into the cavity of the mouth, a portal that allows chemicals to come into contact with the receptors within. Contrary to popular belief, the tongue isn't organised into zones that each specialise in a particular flavour. Instead, each taste bud has a medley of different taste cells nestled within it. But while all parts of the mouth deliver all five taste sensations, some parts are more sensitive to particular stimuli. In particular, the back of the tongue is particularly sensitive to bitterness; if you've been daft enough to ignore the warning signals from your other taste buds to the point where you're about to swallow something that's extremely bitter, the extra sensitivity at the top of the throat might trigger your gag reflex.

The number of taste buds that we each have says something about our evolution as a species, especially when you compare us to other mammals. Generally speaking, carnivores have relatively few taste buds. Dogs have only about a quarter as many as us, while the domestic cat gets by with fewer still, mustering less than 500. By contrast, the mouths of herbivores are loaded with taste buds. Rabbits have around as many as us, while cows have something like 25,000. Perhaps this makes every mouthful of grass a riot of flavour, but the reason for this apparent oversupply of taste buds is primarily about defence against poisons. The ancestors of domestic cattle grazed a broad range of plants, much as wild rabbits and a swathe of other herbivorous mammals do today. The more different types of vegetation that animals encounter, the greater their risk of accidentally eating something that will incapacitate them. Consequently, they need an early warning system, and this is what their extensive battery of taste buds provides for them. Carnivores have far less to worry about in this respect; their diets are far less diverse than their plant-eating cousins. Relatively few of their prey contain toxins, and as a result, evolution hasn't equipped them with such a sophisticated sense of taste. In terms of the numbers of taste buds, we sit somewhere between these two types of diet, perhaps closer to the herbivores than the carnivores, but most similar to omnivores, in particular a kind of mammal with a curly tail and a winsome snout. Our ancestors kind of did eat like pigs, and now we probably experience taste much as they do.

*

As anyone who's had a cold or Covid-19 knows, when smell leaves, it takes a lot of taste with it. If you pinch your nose as you eat, you'll find that the rich experience of foods fades perceptibly. My dad always used to claim that a blindfolded person with a peg on their nose couldn't tell the difference between water and Guinness. In the spirit of enquiry, I gave this a go recently. In the absence of a peg, I clamped my nose with a bulldog clip; in hindsight this was a terrible idea, but I bore the pain in the name of science and got someone to pass me the drinks in turn. With the frothy head taken off the Guinness and both at the same frigid temperature, tasting the difference was far harder than I anticipated. My near taste-blindness under these conditions indicates just how important smell is in our perception of flavour. It's sometimes said that 80 per cent of taste is smell. The values might vary, but the overall message is the same: taste is mostly smell. However, at our current level of understanding, we can't say how much of our ability to perceive flavour can be ascribed to olfaction; even if we could, it depends very much on the food that's being eaten. What we can say is that smell makes a huge contribution to flavour.

For much of the twentieth century, consumers in the UK were treated to commercials featuring the Bisto Kids, a boy and girl who were habitually stupefied by the savoury aroma of meaty gravy. Having smelt the stuff, they'd follow the enticing curls of odour to their source. Given that the Bisto Kids existed during the nadir of British cooking, this was likely to be a sorry dish of overcooked, annihilated vegetables alongside some wizened meat, collectively sinking below the waves of the gravy for the third time. But while a delicious scent is important in priming our appetite, smell contributes most flavour once food is in the mouth. When we think of smelling, it's typically what's known as orthonasal olfaction, which describes air going via the nostrils to reach the olfactory receptors. There is another route to the receptors, which is where aromas rise up from the back of the mouth along the nasopharynx to the olfactory epithelium. Fittingly, since the odours reach the nose from the rear, it's known as retronasal olfaction. The passage from the mouth to the nose is what the great French epicure, Jean Anthelme Brillat-Savarin, described as 'the chimney of the mouth'. As we chew, we break up the food and its aromas flow up to the olfactory receptors; it's by this means that we perceive smells as we eat.

The weird thing is that whether we smell things via the nostrils or the mouth affects our perception. Logically it shouldn't – after all, both activate the same sensory receptors – yet when we smell food, the experience is localised in the nose, but when that food is in the mouth, the experience is centred on the mouth. This illusion explains why we don't seem to distinguish between taste and smell in flavour perception. When we have a sip of wine, we might sense all kinds of flavours, but all that taste is contributing is perhaps a slight sourness, because wine tends to be acidic, and possibly some sweetness. Everything else is coming from the nose – it just seems like the flavour is in the mouth because of a trick played by the brain. It's not entirely an illusion though; scans performed while people are engaged in 'sniff smells' and 'taste smells' show different patterns of brain activation. The market for an air freshener based on durian, cauliflower or blue cheese would be vanishingly small, yet many people love eating them – this despite the fact that the same sense is providing most of the input in both cases. It seems that smell is the only one of our senses that provides two distinct experiences to the same stimulus.

*

To summarise this convoluted picture, when we talk about the taste of something, we don't just mean taste, and when we say that smell plays a dominant role in flavour perception, we don't really mean the kind of smell that we experience when we sniff. The final part of the flavour trinity is chemesthesis, a kind of chemical sensitivity that's similar in some ways to taste, yet distinct. In fact, since the chemicals in question stimulate the nervous pathways that are engaged in tactile, pain and thermal sensations, chemesthesis has most in common with our sense of touch. It's all part of the glorious complexity of this mouth sense.

Two of the chemical agents that we 'taste' via chemesthesis, menthol and capsaicin, not only give us the flavours of mint and spiciness but trigger a response in our thermal receptors that mimics the perception of coolness or heat, without any change in temperature. Menthol causes a short-term change in nerve responses to temperatures so that we become less sensitive to heat and more sensitive to cold, which is why when you breathe in through your mouth after eating a mint sweet, it feels cooler. Capsaicin goes the other way, setting off receptors that would otherwise detect dangerous levels of heat. So chilli heat makes the mouth more sensitive to thermal heat, which can make eating a dish that's both spicy and thermally hot somewhat painful. Eat it often enough, though, and we become less sensitive to it, basically as a result of nerve damage. The nerve damage is reversible, so if you want to maintain a socially impressive chilli tolerance, you need to eat it on a regular basis.

The idea of chemesthesis having more in common with touch than taste is never more aptly demonstrated than in the anguish of someone who's been to the toilet after chopping chillis. Capsaicin isn't the only agent that provides heat, however; mustard, horseradish, black pepper and ginger also provide a warmth that is generated by feel. If you've ever bitten into an unripe banana, you'll be aware of the weird puckering sensation that results. This astringency is caused by a group of plant chemicals known as tannins. Aside from unripe fruit, you can welcome this sensation into your mouth by drinking tea that's been left to brew for too long, or by chewing the skin from a grape. In fact, it's the latter that leads to the mild astringency of red wine. Tannins are alkaline, so taste bitter, yet they also activate touch receptors in the mouth as a result of their interaction with saliva. Similarly, the dissolved carbon dioxide in fizzy drinks reacts with saliva to form carbonic acid and like astringency, this is something we experience by chemesthesis. It wouldn't be unreasonable to imagine that it's the little bubbles of gas that make the tongue sparkle as they burst, it's what most people think, after all. A few years ago, however, scientists went to the length of putting people in a barometric chamber, increasing the pressure within, and then asking them to try fizzy drinks. The pressure stops the bubbles from bursting, ruling out any possible tingle from these, and yet the drink tasted the same. What's actually happening is that the carbonic acid mildly irritates the mouth, which registers the sensation via pain receptors. Irritant or not, try drinking a carbonated beverage when it's gone flat; the joy leaves the drink with the gas and it tastes lousy. I say 'tastes', but of course it isn't a taste – it's a flavour. The distinction is a fine one, but it is important.

There's nothing like a meal to engage the senses. Though the leading players in flavour perception are taste, smell and chemesthesis, our senses of touch, vision, and hearing all have cameo roles. Touch provides us with the awareness of textures, perhaps the sensuous feeling of chocolate as it melts in the mouth or the rubberiness of a cube of *hákarl*. Thermoception promotes the enjoyment of an ice-cold drink on a hot day. Vision calibrates the appeal of food in our brain. And we might hear the appetising crunch of a fresh apple as we bite into it. Eating draws the modalities together in a unique way; it's the most multisensory experience that we can have.

\*

We think of taste as being the preserve of the mouth, yet other animals take a more liberal approach. The bodies of certain catfish, for example, are covered with such huge numbers of taste receptors that they're effectively swimming tongues and, for their sake, we might hope they enjoy their endless smorgasbord of sludge. However, whatever they taste, it won't be sweet – since, like their comic book enemy the cat, they lack the appropriate receptors to detect sugary molecules. Insects have somewhat different receptors to ours; nonetheless, they too have the capacity to taste things beyond the mouth. Butterflies and flies taste with their feet, so are able to register the chemical composition of things they land on. The experience of alighting upon a flower is no doubt a thrill for a butterfly; whether the same can be said for a bluebottle touching down on a turd is anyone's guess. The scourge of many a barbecue, the mosquito, has in its arsenal a blood-specific taste receptor. This is a facet of biology that holds for all of the different senses; animals are shaped by their ecology and their sensory receptors reflect this.

The domain of our sense of taste has been expanding for some time. It was long thought that the tongue was the sole organ of taste, until taste buds were located around the mouth and the top of the throat. Then, more recently, our insides were discovered to be riddled with taste receptors. This story begins in the middle of the last century, when it was determined that the body responds differently to a sugary drink according to whether it enters via the mouth or is injected into the blood. It's only when sugar passes through the gut that the pancreas starts to release insulin. We can't blame the body for responding properly only when we actually eat foods, but the question was, how does it know? The answer, we now understand, is that taste receptors in the intestine tell it. What's more, there are receptors throughout the digestive system, as well as in the lungs, kidney, pancreas, liver, brain and even the testes.

Although these are the same as the taste receptors found in the mouth, they're not organised into taste buds, and they don't interact with our brain in the same way that those in our mouths do. Consequently, we get no sensation of flavour from these receptors, which is just as well since who wants to taste the contents of their own intestines? This, then, isn't taste as we normally understand it; indeed, some of the receptors' roles are downright weird. If we inhale something poisonous, tiny hairs known as cilia that line our airways waft the poisons away, but it's taste receptors in the respiratory system that first detect the toxins. In a similar way, taste receptor cells in the airways, the gut and the bladder detect the presence of invading organisms, such as parasites, based on the tell-tale chemicals that these unwelcome guests excrete. The receptors provide an early warning, alerting the immune system and setting off the compulsion to eject the parasites from the body.

Notwithstanding this amazing diversity of receptor functions, their main job is arguably to detect the presence of nutrients and organise the body's response to them. Taste receptors in the gut and the brain work to monitor energy levels and satiety. When we've not eaten for a while, they trigger the release of a hormone called ghrelin, which is what gives us the feeling of hunger. And if we've eaten a toxin, they put the brakes on the digestive system to slow the passage of the noxious stuff through the body, preventing it reaching the intestines from where it leaches into the blood. The reason that alcohol affects us more rapidly when we haven't eaten is the result of a similar process. When we've 'lined the stomach', the alcohol – strictly speaking, a toxin – gets held up by the stomach for longer before being passed into the intestines.

Taste receptors throughout the body, including the mouth, operate as an incredible sensory network that primes us to seek nutrients that we're lacking. For instance, disease can cause the adrenal glands to produce less cortisol, resulting in low blood pressure. There's a quick, short-term fix for this, which is why people or animals suffering these effects begin to crave salt. It's not something that's consciously thought about, but rather a deep-seated drive. Similar effects can be seen in relation to the parathyroid glands, when the sufferer seeks out calcium, or when there's a drop in blood sugar, which leads to sugar becoming intensely appealing. Many of us will have seen harrowing pictures of children in famine-hit areas with swollen bellies, the result of enlarged livers and fluid retention due to a severe shortage of protein. When food aid is provided, the children bypass all other offerings and go straight for amino acid-enriched soups. Their bodies know where lies the shortage and impel them toward the protein. Thankfully, most of us never encounter such severe nutrient deficits, but nevertheless we're all subconsciously steered by the taste receptors inside us.

\*

Just as with smell, there are huge differences in our ability to taste. The variation is down to genetics and seems to relate partly to the number of taste buds that each of us have. So-called 'supertasters' may have up to four times as many as people at the other end of the scale, who are sometimes dubbed 'non-tasters'. However, the naming is somewhat misleading; non-tasters can usually taste things, even if not as sharply as others. Meanwhile, 'super' implies that having exquisite sensitivity to taste is a good thing, yet it can make mealtimes tricky. Bitter flavours in particular can taste overwhelmingly strong, making many green vegetables, drinks such as beer and even some chocolate treats unpalatable. Here in Sydney, there's no shortage of snobbery about coffee. For many supertasters, however, it's a drink of such unparalleled bitterness that they can't stand it. In fact, only people in the distinctly average taste range find coffee pleasant, something to bear in mind the next time someone becomes too mulish about the choice of cafe.

So how do you know what kind of taster you are, and whether your fussiness has a genetic basis? The simplest way to find out is to pop a bit of blue food colouring on the tip of your tongue, which shows up the tiny bumps on the tongue known as papillae that contain taste buds. It's said that if you can count more than fifteen papillae in a 6mm diameter circle then you're a taster, and if you have more than thirty-five, you're a supertaster. A more scientific approach involves taking a test. Propylthiouracil, or PROP, is a medicine used to treat hyperthyroidism, while its chemical relative phenylthiocarbamide (PTC) inhibits the production of pigments. Neither of them is typically found in the foods that we eat, but both are similar to the chemicals that produce bitter tastes in vegetables. In the context of taste, they generate very different responses when people try them. PTC's properties in this regard were discovered in 1931 by a chemist, Arthur Fox, who, midway through an experiment, accidentally released a billow of fine PTC crystals from his equipment. Though Fox himself didn't sense anything awry, one of his infuriated colleagues complained of a powerful, bitter taste from the spill. Intrigued by the difference in their experiences, and keen to settle the debate between them, Fox decided to test a larger sample of people and found that there really was no common ground: some could detect the taste, others couldn't. We now know that at the heart of this is a single taste receptor gene, TAS2R38, and whether you discern a bitter taste or no taste at all in response to PTC depends on which version of the gene you have. Something like two thirds of people do register a bitter flavour when they try it; since this is a heritable genetic trait, PTC tasting was used as a basic paternity test in the days before more advanced DNA profiling techniques.

Nowadays, PROP is used more often than PTC in tests because the latter is toxic – not an ideal characteristic for taste tests! PROP tests have been used extensively to ferret out who among us is a supertaster. Based on published research, about one person in four finds the taste unbearable, to the point where there are even anecdotes about some supertasters lashing out at the people administering their test and roughly the same proportion find it bland, while the remaining half of people find it slightly bitter, but not overwhelmingly so. It's by this means that people are designated supertasters, non-tasters and tasters, respectively. Although defining people's sense of taste by the reaction of their bitterness receptors to a single chemical might seem a little extreme, I ought to point out that sensitivity to the PROP test tends to correlate strongly with the number of taste buds that people have and with how acutely they perceive other flavours.

A good friend of mine is a supertaster, but for a long time many of her friends simply assumed that she was just picky about foods. Once it was revealed that she's a supertaster, everything fell into place. What I'd taken to be fussiness was actually something beyond her control. She, in common with other supertasters, has a tendency to add dramatic amounts of salt to her food, possibly because of its ability to mask bitterness. This, alongside her disdain for vegetables and leafy greens, represents a risk to her health. Salt has been linked to heart disease, while a lack of vege-tables in the diet increases the risk of some cancers. On the upside, she's no fan of sugary deserts. For supertasters like my friend, the restrictions caused by the enhanced sense of taste mean that a menu can be a minefield; a good diet requires careful planning.

Like supertasters, non-tasters also have their peccadillos. To enliven their comparatively dull palate, they're more likely to reach for the chilli sauce or for highly spiced foods, which are often anathema to supertasters. They tend to be drawn to flavour-ful but unhealthy foods that are high in sugar and fat, the hit that they get from such offerings providing welcome relief from the blandness of other fodder. And in contrast to supertasters, non-tasters are more likely to be able to enjoy tipples like beer and dry wines, while happily tolerating the burn from spirits. The result, perhaps unsurprisingly, is that they tend to weigh more than their supertasting contemporaries and are more prone to alcoholism. Sadly, the bad news doesn't end there. Those who don't have supertasters' – or even tasters' – sensitivity to bitterness are shorn of the oral chemical alarm system for toxins. The body seems to compensate for this by making them more prone to nausea as an alternative line of defence. Unfortunately, this tendency for non-tasters to become nauseous expresses itself in other ways, including a susceptibility to motion sickness. It's important to add, though, these relationships are correlations rather than cast-iron determinations. If you like sprouts or feel queasy in a car, it doesn't mean you're a non-taster.

On a similar theme, no single test can entirely define our flavour perception. After all, the PROP test simply highlights variations in a single gene for one of the twenty-five bitter taste receptors in our mouths. And even though there are correlations between the per-ception of bitterness and taste bud numbers, there's much more to taste than just the receptors. While our taste buds do the detective work, it's the brain that does the heavy lifting when it comes to creating perception. Those regions of the brain that govern our responses to stimuli are more excitable in some people than others, so it's possible that supertasters not only have more receptors but that the parts of their nervous system that interpret the input are more sensitive. This idea comes from studies of the tongue that focus on thermal, rather than chemical, sensitivity. If a small section of the tongue is gradually cooled to around 15°C and then heated with a probe that's a couple of degrees warmer than our body temperature, something weird happens. Around half of all people report a sweet taste, triggered by the change in temperature. Meanwhile, a rapid cooling of the tongue results in a perceived bitterness or sourness. Just as with the PROP test, those people that do experience this strange phenomenon tend to have a much more sensitive response to tastes of all kinds. The overall picture is one of large-scale and broad-ranging variation that emerges from both our hardware and our software: our tongues and our brains.

\*

On the face of it, the data that we have suggests that if we give four people a PROP test, we can expect to get one supertaster, one non-taster and two regular tasters. Whether this actually happens depends on which part of the population you examine. If you've been following the pattern so far in this book, it'll come as no surprise to learn that women have a particularly keen sense of taste. In fact, by some estimates, they are twice as likely as men to be supertasters – even though the genders have very similar underlying taste receptor genes. Why this would be is not well understood. Men and women seem to have similar numbers of taste buds, but women's taste receptors are more responsive, a fact that has been connected with hormonal activity; for instance, oestrogen seems to intensify taste perception. As a result, it's typically women of reproductive age that have the keenest sense of taste. It is at its most acute during pregnancy, when women can have food cravings as well as becoming very highly sensitised to bitterness. A possible reason for this is the critical importance of avoiding potential toxins when the foetus is developing. It might be little consolation to pregnant women forced to give up some of their favourite foods, but their supercharged sense of taste is keeping their baby safe.

Just as sex plays a role in just about all the senses, our age influences how intensely we experience things. As Johann Wolfgang von Goethe glumly reflected: 'One must ask children and birds how cherries and strawberries taste.' Within our taste buds, there's a continual changing of the guard; receptor cells are scrapped and replaced every two weeks, allowing us to maintain peak taste sensitivity. Moreover, our taste apparatus is among the most robust systems in the human body. You can kill it with vindaloo or boiling tea and it'll soon bounce back, ready for more. In fact, you can even remove the surface of the tongue altogether and it'll regenerate. Even so, the number of taste receptors, as well as their sensitivity, drops as we age, to the point where pensioners in particular often suffer the loss of eating pleasure. It's been noted that some elderly people add up to three times as much salt to their food in an effort to make it flavoursome.

At the other end of the scale, foetuses develop a sense of taste just four months after conception. Foetuses voluntarily swallow more amniotic fluid when their mother has eaten sweet foods than when she's had something bitter. Once they're born, babies already possess a full complement of taste sensations, as evidenced by the reports of researchers studying the behaviour of newborns to different flavours. Give them drops of water infused with a sweet or umami tang and they comically smack their lips and even smile. They're not so keen on sour or bitter flavours administered in the same way and vent their feelings by sticking their tongues out and squeezing their eyes shut. Babies' response to salt takes longer to develop – they're neutral about saltiness until the age of about four months, at which point they start to prefer weak salt solutions to water.

The sense of taste continues to develop throughout childhood, with taste buds reaching full size when we're in our mid-teens. Until that time arrives, as any parent will know, children exhibit different tastes to adults. One test involves providing a range of different sugar solutions and asking people to pick their favourite concentration. Adults tend to converge on a sugar level that's close to that found in cola, which is precisely why the soft drink contains the sugar levels that it does. Given free rein to indulge themselves, children pick something almost twice as sweet, which matches the sugar concentrations found in foods that are targeted at them. The reason for this difference is basic biology – children need to find high-energy food sources to supply their rapid metabolism. Where children also differ is in how sugar affects their response to pain. Getting a lolly as a reward for a childhood vaccination is a common ploy in many countries, and there's good evidence to suggest it helps. Experiments show that sugar directly eases their pain in a way that it doesn't in adults. This analgesic effect of sugar also shows up in other mammals, including rats. The surprising thing is that obese children don't get the painkilling or stress-relieving effects from it to the same extent as other children. So in order to pacify themselves, obese children have to up their sugar dose, a pattern that bears the hallmarks of a vicious cycle.

Brussels sprouts and medicines are the stuff of children's nightmares; the root cause is the fact that children are vastly more sensitive to bitterness than adults are. This, alongside children's weakness for sugar, represents a massive problem in the modern diet. By the time kids reach adulthood and their sensitivity to bitterness is diminished, unhealthy food habits have often become firmly established. Moreover, though salt and MSG both mask some elements of bitterness for adults, their subterfuge is far less effective on the alert taste buds of the young. At our current state of knowledge, bribing reluctant kids into tackling broccoli remains the most viable way to address the issue.

Irrespective of differences between us in our tolerance for

bitterness, some vegetables and herbs just taste wrong to some people. For instance, I remember my horror when I tasted fresh coriander for the first time. A friend had invited me to dinner and politeness dictated that I should maintain my equanimity in the face of what was to me the overwhelming soapy flavour of the food. Every mouthful was an ordeal, tasting to me like chicken cooked in the run-off from a hairdressers. It turns out that whether you like coriander depends on which variation you have of a single gene, the catchily-named OR6A2. Something like one in six people share my loathing for this herb, just as a proportion of the population turn their noses up at beetroot; to them, its taste is laden with unpleasant earthy flavours. The culprit this time is the gene OR11A1. Both these genes are connected to smell rather than taste, but each has a powerful effect on a specific flavour.

Ethnicity plays an important role in taste as well. The ratio of one supertaster to two tasters to one non-taster emerges primarily from studies of Caucasians; once research included a more representative range of humanity, things changed. Among East Asians the proportion of non-tasters is around just one in ten, while in Afro-Caribbeans and African Americans it can be as low as one in twenty.

It's hard to explain these patterns, nevertheless some suggestions have emerged. Eating large amounts of bitter plant chemicals can interfere with the workings of the thyroid gland, which produces hormones that play a critical role in regulating the metabolism. The result is that people who eat such a diet have an increased risk of developing goitre, which, while generally mild, can result in reproductive difficulties and occasionally even death. Nowadays, we can treat goitre relatively easily and there's no reason to avoid green veggies. But for the longest periods in human history, the thing that kept us safe from excessive intakes of the bitter plant chemicals that cause goitres was our sensitivity to bitterness. We don't know for sure, but the comparative rarity of bitter non-tasters in some populations (and even the intense perception of bitterness by children) could be connected to this, especially where the local diet historically comprised a high proportion of vegetables.

The idea of diet as a selective force in shaping the genetics of taste is supported by studies examining bitter taste responses in central and west Africa. Starchy vegetables, including cassava, which was introduced to Africa in the sixteenth century, have played an important role in the diet of people living in these regions ever since. The problem is that cassava contains glycosides, which are precursors of cyanide. Careful preparation diminishes their concentration by the time cassava is served, but it's something that our bitterness receptor, T2R16, detects – and it's this receptor that can make cassava taste unpleasant. Perhaps surprisingly, given the potential toxicity of cassava, a high proportion of African people living in cassava-growing areas have a variant of the gene T2R16 that provides them with low sensitivity to the taste of glycosides. This seems strange until it's seen in the context of something much more dangerous than cassava: *Plasmodium*, the parasite that causes malaria. Cassava, and particularly the bitter chemicals it contains, inhibits the development of *Plasmodium* and so protects against the effects of this deadly disease, so the ability to tolerate the bitter tastes that accompany some varieties of cassava can be a lifesaver.

Tracing the genetic basis of our reactions to bitterness is appealing as a subject for research both because of our varying sensitivity to it and the fundamental role it has played in helping us to avoid potential toxins. Our responses to other tastes are typically less extreme, but there are interesting patterns that align with ancestry. People of African or Asian descent seem to perceive a broad range of different tastes more intensely than Caucasians, a finding that agrees with studies of ethnicity and bitterness tasting.

Another avenue of research has examined how we respond to sweetness. Far from being universally enthusiastic in our response, we vary quite a bit in this respect. Some people can't get enough of it, while others are far less keen. There are people who fit these categories in every population, but there is an overall trend for East Asian people to be less fond of sugar than people of European or African ancestry. Like any such trends, it raises the question of how far our taste is down to genetics and how far it reflects the culture that we grow up in. The answer is that it's a pretty even split, with a fair portion inherited from your parents, but a slightly larger component coming from the society around you. This cultural element is one that concerns medical professionals, particularly in the West where tastes and diets have changed more in the last two generations than at any other point in history.

*

In 1860, an ambitious expedition left Melbourne to explore the Australian hinterland and ended up in folklore as a tragic monument to failure. The goal was to travel north for 3,500 km, mapping the interior which at the time was largely unknown to European settlers, and plotting a route for the telegraph that was needed to connect Melbourne, then the British Empire's second largest city by some estimates, to the rest of the world. It was led by an Irishman called Robert O'Hara Burke, a former soldier who was working as a policeman in the city. A strange choice to lead such a trek, Burke was famed for his complete inability to navigate. It was said that he often got lost between the pub and his home. Nonetheless, it was a huge event. Fifteen thousand people gathered to watch the party as they departed with their horses, camels and wagons on 20 August, carrying 20 tonnes of equipment, including such essentials as a cedar wood dining table and chairs, a bathtub, and a giant Chinese gong. Half of the wagons broke before they left Melbourne. One of these didn't even make it further than the edge of the park where the party were given their send-off. Stoically, if ill-advisedly, Burke and his men carried on.

The first weeks of the trek saw no upturn in their fortunes.

Their progress was hampered by heavy rains and muddy tracks, leading Burke to order that much of their travelling equipment be left behind. Not only that, but they ditched vital provisions, including their lime juice and sugar. The final straw for the expedition's second-in-command, George Landells, came when Burke monstrously insisted on leaving behind a barrel of rum. Landells' departure was only one of many, as Burke's tempestuous nature led to over half the party resigning or being fired by the time they reached Menindee, a tiny settlement on the Darling River only 750 km from Melbourne.

Already, they were weeks behind schedule, condemning them to travel the most arduous stretch of the journey in the heat of the Australian summer. Burke continued to thin the party both by design and by dint of his quick temper, until only four remained with half of the journey still ahead of them. Remarkably, despite the ever-shortening rations and soaring temperatures, they made it to the Gulf of Carpentaria on the north coast of Australia in February 1861. Although they'd achieved their objective, they now faced an even greater problem: the return journey, with their rations almost exhausted.

On that grim retreat, they were forced to sacrifice their remaining animals for meat, but it wasn't enough. Gradually, the travellers succumbed to scurvy, beri-beri and dysentery, brought about by malnutrition. In the end, only one man, John King, made it back alive. He survived thanks to his relationship with the indigenous Aboriginal people, something that Burke and the other members of the group couldn't bring themselves to do. The Aboriginal knowledge of country has developed over tens of thousands of years of living in the very areas that the expedition crossed. Far from being the barren, scorched waste that many imagine, the Australian bush has abundant natural resources. It offers a rich larder to those who know how to find it. In effect, Burke and the others starved amid plenty. Some of the foods that the Aboriginal people could offer, such as witchetty grubs or bogong moths, lie outside the comfort of those not familiar with them, yet they provide vital calories as well as being excellent sources of the very vitamins that the expeditioners lacked. It's hard to avoid the conclusion that they died through a mix of naivety and squeamishness.

In an Australian supermarket, you might imagine that you'd find all manner of interesting foods in the fresh produce aisle. Australia is home to some incredible foods – bunya nuts, finger limes, quandongs, pepperberries, lemon myrtle – all of which have played an important role in the diet of indigenous people for millennia. They're also well suited to growing in the Australian climate. Yet instead of harnessing the natural tastes of Australia, European settlers simply transplanted their own way of life and with it, their food preferences. They cropped wheat, raised cattle and cultivated apples. Later, Asian immigrants contributed their own influences, and now bok choi, bitter melons and – craziest of all in an arid country – rice is grown. The result is that native Australian produce has largely been spurned. This, alongside the dominance of long-established crops, is a strong signal of the importance of culture and tradition in the human diet, which has a limiting effect on the breadth of our menus. Research from the University of Washington suggests that most people eat no more than thirty different foods on a regular basis. The question is what, if anything, does taste have to do with this?

Across the world, and regardless of culture, infants are pre-programmed to have pro-sugar, anti-bitter taste preferences. But these innate inclinations are moulded by experience. We start learning about flavours early – before we're born, in fact. A remarkable study carried out in France described how the foods eaten by pregnant women in the last couple of weeks before giving birth help to shape the preferences of their babies once they're born. In this case, the women ate anise-flavoured sweets and cookies. When offered a sniff of anise at the tender age of just three hours, the infants of anise-eating mothers responded by attempting to suckle, while those of anise-avoiding mums pulled a face. This 'mother knows best' approach continues during breast-feeding. All manner of flavours, from garlic and mint to carrot and cheese, come through loud and clear in breast milk and guide the responses of babies in favour of the familiar. Breast milk isn't the only pointer for infants, though. Those given formula also learn from the experience, and interesting differences develop between babies that are fed cow's milk formula and those raised on soy-based or hydrolysed protein* milks. Compared to cow's milk, these latter two types have a spectrum of flavours that includes some bitter notes and a more savoury flavour. The result is that such infants tend to accept the transition to savoury foods better than their cow's milk formula-fed peers and, years later in child-hood, were also happier to drink sour apple juice and to eat the dreaded broccoli.

Once infants are weaned onto solid foods, they're quick to show their disapproval in the face of unsatisfactory offerings. A steadfast refusal to entertain vegetables is common, and the sight of sundry greens being jettisoned from the highchair can bring tired parents to the brink of despair. Nonetheless, persistence is the key to developing good eating habits, especially during a particularly flexible period in a child's development, between four months and two years of age. There's a huge number of studies supporting the idea that repeated exposure to unfavoured items can make a solid dent in a youngster's resolve, and a whole swathe of veggies and fruits have been used to test the mettle of recalcitrant children. The consistent finding is that a few days of bloody-minded perse-verance on the part of the parents starts to turn the tide. Though kids still might not greet their green nemesis with delight, they increasingly begin to humour their feeders. With a little luck, a kind of foody Stockholm syndrome might eventually develop and junior moves from tolerance to liking. It can be a battle of wills,

---

* This is sometimes provided to infants who struggle to digest complex proteins.

the first generational conflict between parent and offspring, but if you find yourself in this position, take heart: the research also shows that the food habits that become established early on tend to persist into later childhood and beyond.

Relative to other animals, humans spend an incredibly long time developing to independence and, for most of this time, children eat the food that their parents provide them with. So the flavour preferences of the parental generation are passed on to a captive audience of kids, and cultural predilections become firmly established, which is one reason, for instance, that the Danes – the world's greatest lovers of cheese – eat a staggering 28 kilograms of cheese per person each year, while the Chinese restrict themselves to just 100 grams. It also explains regional fondness for things like durian in South Asia, fermented herring in Sweden, or sour tamarind in India. It also gives insight into why food manufacturers tailor their products according to which country they're selling it in. The KitKat, one of the world's bestselling chocolate bars, comes in all manner of flavours in Japan, such as green tea, while there's a hazelnut-inspired offering in praline-loving continental Europe. At the same time, the same food can have varying interpretations in different parts of the world. In the West, vanilla is identified as being sweet, based on its association with desserts and puddings, while in East Asia it's added primarily to savoury dishes and isn't thought of as being sweet at all.

These cultural differences go beyond a simple preference for certain foods and condition our brains in astonishing ways. An intriguing experiment carried out in 2000 by Pamela Dalton and her colleagues at the Monell Chemical Senses Center in Philadelphia demonstrated that what we experience as flavour is guided by our expectations. Dalton asked American volunteers to sniff benzaldehyde, a scent that's redolent of the flavour of almonds and cherries, and that – from a Western perspective – is associated with sweet treats. However, the concentration of benzaldehyde provided was weak, giving them only the vaguest impression of a smell. At the same time, the volunteers were given a drop of either pure water, water with a tiny amount of saccharin, or water with an equally tiny measure of MSG. With just water, or 'savoury' MSG water in their mouths, the volunteers responded fairly neutrally to the benzaldehyde. But with just a minimal amount of sweetness on the tongue, the benzaldehyde smell came into sharp focus. Essentially, their brains had made the link between the two associated bits of information and delivered a clear perception. Even more fascinatingly, when the study was repeated using Japanese subjects, a different pattern emerged. In this case, the benzaldehyde smell was brought to life by the MSG water, most likely reflecting the fact that, in Japan, almonds are typically thought of as savoury rather than sweet. We know that certain neurons in the brain respond to particular couplings of smells and tastes. What's more, these flavour associations develop through the experiences we pick up in our lives, demonstrating the importance of culture in shaping our perceptions of flavour.

\*

Although our cultural preferences and predilections form the foundations of our tastes, we continue to learn about flavours throughout our lives. As we reach adulthood, we might start to appreciate more robust flavours, like strong cheeses or red wine, and we become more exploratory. Beyond middle age however, we start to become hidebound. Studies of people moving between countries show that those who arrive in a different culture after the age of around forty tend to stick to what they know. Whenever we sample some new food, the immediate taste sensation in the mouth is backed up by metabolic feedback from the rest of the body as we digest it, which shapes our response. The strongest signals occur when we eat or drink something that tastes pleasant but which we react badly to. For example, I can't go near tequila after a slight overindulgence some years ago – my brain learned the consequence of too much tequila and didn't like it one bit. We learn things that we like in a similar way, albeit the process is more subtle. Information on both the flavour and the metabolic consequences of a food are integrated by the brain, triggering reward networks such as the mesolimbic pathway, and promoting a feeling of well-being when we eat something that we like. It does this by releasing dopamine, a neurotransmitter that in turn affects our behaviour. In the case of foods, it means that the brain acts as the arbiter of our preferences, regulating the likelihood that we'll go for that food again.

Creating the perfect experience for the palate is a question of balance. Take sugar, for instance. Our liking for sweet things is a matter of degree. Add sugar to something and we enjoy it more, but beyond a certain point further sweetness becomes repellent. Though it varies from person to person, we seem to like solutions that comprise around 10 per cent sugar, which is sometimes referred to as the 'hedonic breakpoint'. Salt also has a breakpoint, which is around 0.5 per cent. For the most part, our breakpoints are fairly consistent, but context does play a role. We're happy to put up with sourness in yoghurt, for instance, but in milk, the same level of sourness would likely make us pour it down the sink. Since virtually all foods comprise a mix of different flavours and induce contrasting responses in the body, we sometimes end up taking the rough with the smooth. The bitter elements in drinks like coffee or beer are presented alongside any nutrients we might choose to add, like sugar or milk, and also with chemicals, caffeine or alcohol, that give a buzz. In some ways, we accept the less pleasant aspects of the taste because they're offset by things we like. There are interactions, too, when there's a mix of things we do like. Fat and salt often coexist in the modern diet, especially in fast foods, and there's some evidence to suggest that fat plays a role in dampening our perception of salt, which means, naturally, that you have to add more salt to get the same effect.

It's this interaction between different tastes that has been the focus of some of the most intense research carried out into the human senses. Taste, and in particular the way it translates complex chemicals into flavour sensations for our hungry brains, is huge business. Millions of years of evolution have provided us with the equipment, in the shape of taste buds, to seek out especially rewarding foods. Sugars, fats and salts thrill our tongues because for most of our history they've been rare. Now, however, they're cheap and easily accessible, yet the brain still treats them as exotic and very welcome visitors. As we eat foods laden with them, our brains release endorphins and create links, associating the pleasure of eating with a particular food and encouraging us to indulge again and again. Food manufacturers realised long ago that if they could harness this powerful inbuilt mechanism, the commercial opportunities would be huge.

They set to work, targeting the brain's pleasure systems with surgical precision. The flavours, and especially the balance of sugar, salt and fat, are engineered to what Howard Moskowitz, an American psychophysicist and market researcher, referred to as the 'bliss point', at which the flavour sensation is optimised. While natural foods often contain these elements, they aren't tailored in the way that fast foods are, and consequently don't give us that same instant hit. It's this hit that keeps the tills full at burger joints and their like around the world.

In Roald Dahl's masterpiece, *Charlie and the Chocolate Factory*, the enigmatic Willy Wonka is obsessed by his quest to develop the most incredible, intense experiences and flavours, like the Whipple-Scrumptious Fudgemallow Delight or the Scrumdiddlyumptious Bar. This isn't what real fast-food manufacturers look for. Research tells them that their products' taste should be neither too strong, nor too weak: enough to pique our interest, but not so much that it overwhelms with its complexity. In this way, fast food delivers an immediate yet transient sensory buzz.

The other crucial aspect of fast food's armoury is 'mouth feel', the sensation we get from eating it. For instance, the ideal French fry has a crunchy outside and a soft centre – it's this contrast in textures that delivers maximum enjoyment. Emulsified foods spread the joy around our mouths to excite all of our taste buds. Eating chocolate is so pleasurable because, with a melting point around 36°C, a fraction below our body temperature, it melts in our mouths as we eat it. By the same token, the texture of fast food shouldn't be too robust. The psychology behind this suggests that if a food is rapidly processed as we eat, without requiring too much chewing effort, the brain registers it as being relatively low in energy and keeps us coming back for more. By catering so meticulously to these predispositions, the fast-food industry has discovered a kind of trip switch. The result is, of course, lots of repeat custom and the Western world's obesity crisis.

Our proclivity toward fast food has another insidious effect. Sensory systems are very often quick to adapt, which in the context of food means that if your diet is high in sugar, salt or fat, you respond to these changes. First, the receptors become less sensitive, and second, the genes that code for the taste receptors are downregulated. What this means is that, for example, the more salt you have in your diet, the less keenly you taste it. Consequently, you add more salt, because the alternative tastes less appealing, and the process continues. This is a temporary change and entirely reversible; if you restrict one of the key flavours from your diet, you'll start to become more sensitive to it. One study, using twins to control for genetic differences, showed that the genes with the blueprint for fat-taste receptors become more active when we're on a low-fat diet. This is, in a way, our tastes increasing their diligence in seeking out a key nutrient. Similar things happen with sugar and salt, but one taste where our perception seems to be very different is bitterness. When we eat a large amount of plant-based foods, we're potentially eating a large number of bitter phytochemicals. In this case, rather than cutting back on bitter receptors, the body responds by increasing their production, which again demonstrates the importance of taste as the gatekeeper against potentially dangerous plant chemicals.

*

While we might imagine taste and smell to be the dominant players on the sensory set of the palate, things can go awry if all our senses aren't in alignment. The Roman gourmet Apicius noted almost 2,000 years ago that we eat first with our eyes, and this goes far beyond the sumptuous plates presented by competitive Instagrammers. For one thing, intensely coloured foods lead us to report correspondingly intense flavours. For another, when visual and taste pairings match our assumptions, our perception of flavour is enhanced. People describe red strawberry smoothies being far tastier and sweeter than the same smoothie dyed green, even though the contents are identical in every other way. When vision and taste don't match up, strange things start to happen. In most cases, what the eye sees, the tongue tastes; if people are given a lime-flavoured drink that has been coloured red, they often describe it as tasting of cherry.

The most famous demonstration of this sensory oddity came in 2001, when Frédéric Brochet, a PhD candidate at the University of Bordeaux, tricked a large group of wine experts by adding red food dye to white wine. His experiment was run across two tasting sessions. In the first, he provided his victims with a glass of red wine and a glass of white and asked them to describe them. Obligingly, they characterised the red wine using words including 'cherry', 'raspberry' and 'intense', while reserving terms such as 'floral', 'crisp' and 'lemon' for the white. Some days later, the experts were invited back, and this time they were given two white wines, one of which had been doctored to appear red. When they were invited to break out some adjectives, the overwhelming majority described the trick red wine in similar terms to that which they'd previously employed for the real red. Although the experts were doubtless embarrassed by their faux pas, Brochet's motivation wasn't to make fools of the experts, but to probe the basis of our perception. As a postscript to the story, he later left academia to pursue a career in – what else? – winemaking.

The ability of visual cues to prime our expectations about food and to influence our perceptions is powerful. Even before eating chocolate, for example, the brain makes assumptions based on its shape. Round chocolates conjure the idea of smoothness, which leads to an assumption that the flavour will be sweeter, creamier and less bitter. In contrast, angular shapes inspire thoughts of complexity and even harshness. This is a form of what's known as cross-modal association; we assess the quality of an object using one sense and then interpret what that quality might mean for our other senses. It's one possible reason that square dinner plates have never really caught on. We know that beliefs predispose us to affect our enjoyment of things. For instance, people enjoy Coca-Cola more when they drink it from a branded Coca-Cola cup, and they prefer drinking wine when they believe it to be expensive. This being the case, how do we separate visual cues from preconceptions?

Some years ago, a new approach to eating was developed in restaurants around the world. The idea was to decouple the dominant sense of vision from the experience of eating and allow diners to concentrate more directly on flavour. Le Gout du Noir ('the taste of black') was opened in Paris in 1997 and was quickly followed by similar enterprises throughout Europe and North America. Customers either ate in a completely darkened restaurant or were provided with blindfolds. Did it work? Perhaps unsurprisingly, opinions were split. Some people reported that their perception of flavours was intensified, while others thought it was no more than a gimmick. When a test was conducted under controlled conditions in a lab in Germany, the results were clearer. Eating in the dark resulted in people enjoying their food slightly less and being more unsure about what they were eating. Another surprising outcome was that blindfolded people ate far less than usual, yet assumed they'd eaten far more than they actually had. The overall conclusion is that though such an experience has lots to recommend it in terms of novelty, vision does make an important contribution to the pleasure of eating.

When it comes to wine-tasting, our ears also get in on the act. Changes in the ambience of the occasion, particularly the lighting and background music, profoundly change the tasters' assessment of the wine they're drinking. Lively, higher-pitched music tends to increase the perception of acidity, while more mellow tunes accentuate wine's fruitiness. In other words, the music helps to set our expectations for the way that the wine will taste. Based on these presumptions, taste falls into line. One of the world's leading authorities on cross-modal interactions like these is Charles Spence, of Oxford University. In 2014, Spence ran an event at a London food and drink festival to explore the effects of music and lighting on people's enjoyment of wine. Volunteers were invited into a windowless room and presented with some rioja in a black glass. Next, they were asked to assess the wine while the music and lighting in the room changed. Rather than the taste remaining constant, which is what we might imagine, it altered immediately according to the conditions. Red lighting, for instance, seems to have the effect of making wines taste sweeter, while green and blue hues convey spiciness or make them seem fruitier.

Heston Blumenthal, the gastronome and celebrity chef, is well known both for his eccentric and ground-breaking cuisine, and his attention to sensory detail. His 'Sounds of the Sea' dish recreates a beach in edible form, complete with delicious imitations of seaweed and sand. Along with the food, diners are presented with an iPod that plays seaside sounds. Although some may be quick to dismiss this, it's based on research carried out by Spence which reveals that people enjoy their food more against the backdrop of a marine soundscape. Perhaps the most surprising development in this context came when Blumenthal was devising the dish that he's perhaps most famous for: egg and bacon ice cream. The initial results of his experiments weren't promising, largely because the egg and bacon flavours merged into one another. But, as Spence relates, the crucial development was to add a piece of crispy, fried bread. The crunchiness of the bread seemed to magically separate the bacon flavour from the egg and so helped to propel the dish to fame.

For all that we are taught at school that our senses are entirely distinct from one another, there's a huge amount of sensory cross-talk. This is arguably at its most extreme in the case of taste, which, though synonymous with flavour perception, invokes the most remarkable sensory collaboration that we experience in our lives.

# Skin Sense

*Touch is ten times stronger than verbal or emotional contact, and it affects damn near everything we do. No other sense can arouse you like touch. We forget that touch is not only basic to our species, but the key to it.*

*– Ashley Montagu in* Touching

A human foetus, at eight weeks, is roughly the size of a kidney bean. We can see the beginnings of facial features, and it's beginning to look recognisably human. It still has a long way to go, yet the tiny body is permeated by a network of nerves. It is also developing its first major sense; it is beginning to feel. Six weeks later, the foetus is now the size of a lemon. Its spindly limbs are usually folded in toward the body, but occasionally tiny hands and feet reach out to their surroundings. Miniature fingers grasp the umbilical cord and touch the walls of the uterus. The foetus explores its own body and occasionally sucks its thumb. Twins, sharing the same closely confined space, have something extra to discover, and even at this early stage, they're touching each other. Another month on and they've become fascinated, spending almost a third of their time touching their sibling, putting much more effort into investigating their twin than themselves.

The extent to which twins interact through touch, even before they're born, shows how instrumental touch is to our lives. This inquisitiveness continues after birth, as the baby delves into a whole new universe of touch sensations. As she handles objects, feeling their shape and texture, she's building new connections among the neurones in the brain. This isn't solely about touch alone, however. As the baby reaches for the objects around her, she's using touch to lead the way in developing and integrating other senses, especially vision. As she stretches and grasps at a brightly coloured block lying on the floor near her, she's calibrating her vision, getting a feel for operating in a 3D world and developing the spatial awareness that will be so vital in later life. In this way, we all feel our way through our early days and weeks, using touch as our pathfinder sense.

As babies explore their world, they're using discriminative touch, actively finding out about the shapes and textures that they come into contact with. This is touch at its most practical, but in the past couple of decades we've begun to recognise that there's another aspect to this sense, one that we call emotional touch. We've always known that hugs and caresses were important, but we hadn't fully realised that the two types of touch have a separate nervous architecture. Discriminative touch is wired to reach the brain rapidly, along high-speed nerves known as Type A fibres. By contrast, emotional touch dawdles, travelling up to fifty times slower along C-Tactile fibres, the country lanes of our nervous system.

When signals from the two types of touch reach the brain, they're treated differently to one another; distinct neural networks are activated to process the incoming impulses. The result is a two-tier touch arrangement: a rapid response system that allows us to collect information about the world, and a slower secondary network that comes into play when we are touched. Essential though discriminative touch is, we've long underappreciated the body's dedication to emotional touch, even though, according to some estimates, we have something like three times as many nerve fibres devoted to it.

Emotional touch provides the foundation for our social

tendencies. Touching, and being touched by others, is profoundly important to us from the moment that this sense develops in the womb. We greet one another with a fist bump, a handshake or an embrace. We encourage with a pat on the back, or console with a cuddle. We affirm our love for our partners with tender kisses. As we engage in all of this, the brain responds, catalysing the release of endorphins, oxytocin or adrenaline, thereby priming both our mood and our behaviour. It's this fundamental connection between body and mind that places touch at the heart of our social interactions; its immediacy and intimacy provide the foundation of our relationships.

Long before we evolved spoken language, touch was a primary means by which we communicated and it remains important to this day. In a series of experiments investigating how accurately we can represent our feelings purely by tactile means, participants were asked to touch a blindfolded stranger briefly and in a way that conveyed a particular emotion. The stranger then had to decide what was being communicated. The message was decoded incredibly effectively, especially for emotions such as anger, fear, love, sympathy and gratitude, the affective compass points of our social interactions. We may not think of touch as having a distinct lexicon of terms, but we're so attuned to it that we all intuitively speak its language. Though we certainly underrate touch, especially compared to senses like vision and hearing, it is a sense with ancient evolutionary roots that provides the basis for our rapport with others and our ability to relate to our surroundings.

We can recognise the importance of touch even in the humblest of animals. The roundworm is a model species in biology, which is to say that it's a scientific poster child and a standard that's studied extensively as a means to try and unpick the mysteries of life. The Worm, as it is known to its champions, is a simple animal with a near transparent body and only around 1,000 cells, compared to perhaps 40 trillion that comprise you or me. In laboratories worldwide, these creatures, barely larger than an iron filing, exhibit what we might politely call a limited behavioural repertoire. Yet even in this most unassuming of animals, the stimulation provided by touch is essential. If a single worm is reared without the ability to physically interact with other worms, it grows more slowly and fails to learn how to respond to danger. Patient scientists, exploring the deep origins of the senses, examined whether this could be remediated artificially and found that gently tapping the lonely worms with a probe helped to restore their proper developmental path.

A little closer to home, studies on rats have shown the importance of maternal love and attention to the healthy development of pups. Baby rats who are lavished with affection, particularly being groomed and caressed by their mothers, develop into well-adjusted adults, while those who suffer neglect become nervous and edgy. The shunned pups chase their tails, are prone to over-eating, and struggle to interact with other rats. Tellingly, they become less attentive parents themselves, so that the problem echoes through the generations. The question is, is the deprivation of touch responsible for all this? As with the worm, it required a little direct human intervention to find out. Armed with warm, damp paintbrushes, loosely mimicking the tongue of a mother rat, researchers carefully stroked neglected newborn pups, an apparently trivial intervention that was enough to ameliorate the worst effects brought on by poor rat mothering.

Our closest ape relatives – chimpanzees and bonobos – dedicate as many as five hours each day carefully stroking, brushing and picking at one another. You might imagine that this says something about their fastidiousness in the face of plagues of fleas and other blood-sucking nasties, but if you did, you'd be wrong. Nor is it a matter of vanity. While this kind of thorough grooming undoubtedly helps maintain both the primates' hygiene and their immaculate appearance, the importance of grooming is in building and maintaining social bonds. As they caress one another, they stimulate the release of oxytocin, sometimes called the love hormone, which in turn promotes sociability. Grooming is the glue that binds ape society together.

This kind of social touching is no less important in humans, but it's something that we devote far less attention to than our animal relatives. Even if we have an idea of how rewarding touch can be, many of us are touch-timid, or at best touch-tentative. The Covid-19 pandemic has increased the isolation that we each feel to the point that interpersonal touch seems to have become, as Shakespeare said, 'a custom more honoured in the breach than the observance'. We might turn to history to find some examples of just how important touch is, and how disastrous it might be if we lose this aspect of our lives permanently.

*

In the thirteenth century, the Holy Roman Emperor Frederick II was a man whose thirst for knowledge substantially exceeded his moral scruples. In between governing his diverse territories, Frederick was exercised by the question of what language was spoken by Adam and Eve. As far as he was concerned, it had to be either Hebrew, Greek, Arabic or Latin but to narrow his shortlist, a little callousness was in order. He decreed that a number of babies be taken from their mothers and reared under strictly controlled conditions. In particular, he insisted that the infants were raised in silence so that, when they started to speak, they'd fall back on their innate language. Consequently, he employed nurses but prohibited them from speaking to their charges. Even more troublingly, he forbade the carers from engaging in anything but the most basic physical contact. The experiment did produce a dramatic finding, though it was entirely unconnected to the original question. Although they were well fed and washed, the absence of touch, of physical affection, was catastrophic. Shorn of human contact, one by one the babies sickened and died.

The not-so-holy Holy Roman Emperor's experiments have

echoes in modern times, infamously in the horrendous cases of neglect that were uncovered after the fall of Romanian dictator, Nicolae Ceauşescu, in 1989. Ceauşescu's policies, in particular his banning of all forms of contraception and the tax that he extracted from childless people, helped to drive a huge increase in the birth rate and a corresponding influx of children into state orphanages. For many of these children, the conditions in their institutions were brutal and inhumane. They faced severe beatings on a regular basis to force them into mute compliance. Arguably worse was the neglect. It may not leave visible bruises, but being starved of attention, love and stimulation can be felt even more deeply.

Medics and humanitarian workers who visited the facilities in the aftermath of Ceauşescu's overthrow described how toddlers would reach out to them through the bars of cots that had been their *de facto* prison cells. Many of them had never experienced tenderness in any form and seemed to crave it. Yet if the children were picked up, they'd resist the hug, only to demand another the moment they were put down. The basic interplay of a human embrace seemed unfathomable to them. Even once they'd been extricated from their hellish conditions, many struggled to bond with other people. The anatomy of their brains had been sculpted by their experiences, and the emotional connections that form the bedrock of our lives were simply lacking. The years since have brought about many improvements for the Romanian orphans, most of whom are now of course in middle age, but the scars of neglect may never fully heal.

A decade or so before the world learned of the horrors in Ceauşescu's orphanages, a crisis of a different kind was developing in Colombia. In the maternity unit of San Juan de Dios Hospital in Bogotá, paediatricians were struggling to cope with the number of pre-term babies that were being delivered. Around one in every eight births worldwide is premature, meaning that it occurs at least three weeks in advance of the due date. The combination of low weight, incomplete development and their struggle to maintain the right body temperature places pre-term babies at disproportionate risk. Medical interventions, particularly in the form of incubators, help enormously, but the problem faced in Bogotá by doctors Edgar Rey and Hector Martinez was that there weren't enough of these machines to go around.

According to legend, Rey came across an article about how kangaroos give birth to delicate, grape-sized young that remain in their mother's pouch until they're strong and large enough to face the world. Rey was quick to see the parallels with human pre-term babies and it was this that provided the inspiration to develop what became known as Kangaroo Care. Rey and Martinez devised a simple but highly effective solution to the problems caused by the shortage of incubators. The babies would be wrapped tightly against their mother's chest, skin to skin. The immediate benefit of such close contact was that newborns stayed warm, but the benefits of Kangaroo Care extended far beyond mere temperature.

We now know that direct contact between mother and baby results in an impressive catalogue of benefits for the newborn. Their sleep patterns improve, they begin breastfeeding sooner and gain weight quicker. Perhaps unsurprisingly, they cry less. The simple act of being held and cuddled by mum* causes the levels of a key stress hormone, cortisol, to drop. In turn, the mother benefits in precisely the same way.

Babies don't have to be fastened to a parent to benefit. Even a few minutes of dedicated caressing each day can yield impressive results. A study by Tiffany Field of the University of Miami showed that pre-term babies who received tactile therapy in their first days gained almost 50 per cent more weight again as those who'd missed out. Moreover, touch has the power to ameliorate one of the most concerning issues faced by premature babies,

---

* There's plenty of evidence that fathers can get in on the action too, it's just not as extensively studied.

namely that their early entrance to the world can disrupt the development of their brain. Imaging studies have revealed an increased activation of crucial brain regions in babies undergoing touch therapy, and the effects seem to extend beyond those precarious early days of life. At the age of ten, children who benefitted from skin-to-skin contact as babies deal more effectively with stress, sleep better and demonstrate all manner of cognitive improvements.

Given the chance, babies naturally let their parents know how a little touch can help. A study that contrasted the behaviour of infants who were either cooed and grinned at by an adult or received the same treatment while also having their legs and feet stroked provided clear results. When touch was added into the equation, the babies smiled and babbled more, and they cried far less. So while pulling faces and chattering to a baby is sometimes our default option, it's touch that supplies the magic. This represents a contrast with how we most often interact as adults. And although many of us may be quick to instinctively give a hug or a cuddle, we've until recently been slow to understand the deeper significance of touch.

\*

Since we've made a distinction between two kinds of touch, it's important to point out that we have two different types of skin. Most of our bodies are covered in so-called hairy skin. To be clear, I'm not implying that anyone reading this has a pelt like a woodland beast – hairy skin is simply that which carries hair follicles. The hairs are often barely noticeable. Our legs, torso, arms and head are covered in this kind of skin, leaving just a few places, such as the palms and soles of our feet, our nipples and parts of our sex organs for the other type, which is known as glabrous – meaning hairless – skin.

Each type of skin plays a particular role in our sense of touch.

Glabrous skin is richly supplied with highly sensitive receptors; this is why we actively use our hands to explore through touch and it's also why sex is fun. Hairy skin, by contrast, is far less often used to actively feel objects. Instead, it is primarily dedicated to letting us know when something comes into contact with us. It even allows us to feel insubstantial things, such as the waft of a soft breeze, which blows the tiny hairs on our faces and excites nerve endings in the hair follicles themselves.

If hairy skin isn't so valuable in discriminative touch, it nonetheless responds strongly to being touched. A gentle stroke from a loved one on our neck or back isn't merely a peripheral experience, it's one that our bodies are primed for. Receptors in our hairy skin are particularly sensitive to dynamic stimulation, meaning that they're enlivened by movement. Using robots armed with paint brushes to stroke a series of volunteers, researchers have revealed that there's a perfect stroking speed for maximum pleasure. Too slow or too fast and we don't register the sensation as being pleasurable, but if we're stroked at a speed of 3 to 5 centimetres per second* – even by a machine – the vast majority of us are enraptured. Our heart rate slows down, our blood pressure drops, and we relax. Simultaneously, our brains release natural painkillers and opioids that make us feel warm and fuzzy.

The effects can extend to longer timescales, too. Those of us who enjoy close physical contact not only feel happier, but we tend to be healthier as well, and have stronger immune systems. Tactile interactions come in all sorts of forms; from a loving embrace to colliding with someone who's watching their phone as they step out of a lift. Accidental contact aside, as a means of building a bond rapidly, touch takes some beating. Momentary contact with a stranger can make us warm to them in unexpected ways. It's long been known that if a person touches us on the arm, we're more likely to comply with whatever they ask. If you're rebelling

---

* Some sources expand this to 1–10 cm/s, so there's quite a margin of error.

at the idea of this, you're not alone. Around one in four of us dislikes being touched, but for the rest of us the data are clear. In all kinds of scenarios, from being stopped in the street by a person conducting a survey, to entering a shop and being greeted by a store employee, a fleeting physical connection is like a key that unlocks the warmer side of us.

In a simple experiment in French restaurants, a brief touch on the forearm from the server resulted in diners reporting greater enjoyment of their meal and giving larger tips than those who weren't given such a personalised experience. That said, the same touch that feels pleasurable when we're relaxed can easily become irritating when we're cross – waiters looking to increase their tips might be well advised to judge the situation carefully before handing out carte-blanche caresses.

The phenomenon extends to the world of sport. In volleyball or doubles tennis, team members indulge in a colossal amount of touching at the end of every point. It might be a triumphal high five or a subdued embrace, depending on whether a rally was won or lost. It might even extend to a chest bump or some other kind of enthusiastic freestyling contact. To a restrained English viewer, it all looks a little excessive, but the science tells us otherwise. Back in 2008, Michael Kraus and colleagues from the University of California, Berkeley, carefully measured the amount of touching that went on among team-mates in an early season NBA basketball match. Then, controlling for other variables, such as how good the team were, they examined how well the frequency of touching in that match predicted the success of the individuals and their team across the season. Remarkably, the relationship was strong; more touching meant more success.

Why would this be? It seems that touch is the indispensable catalyst for relationship building. It makes us warm to a person and more likely to co-operate with them. At the same time it calms stress, potentially easing anxiety about performance. The insular cortex in the brain plays a vital role in sensory processing as well as in shaping our emotions. When we touch, or are touched, the insular cortex is activated. If the tactile gesture is appropriate, this helps to shape our emotions positively, strengthening bonds and building trust.

A famous study from a few years ago examined the responses of married women who'd volunteered to be given an electric shock in the name of science. While they underwent this procedure, their hand was held by either their husband, an unknown male researcher, or no one at all. At the same time, the women were wired up to brain imaging technology to see how effectively they dealt with both the fear of the imminent shock and the unpleasantness that attended its subsequent delivery. It turned out that handholding toned down the women's anxiety levels dramatically, especially when their husbands were doing the job. What's more, the quality of the relationship of the couples was crucial; the closer they felt to each other, the stronger the effect. So touch doesn't necessarily provide benefits indiscriminately – it depends on who's doing the touching.

This paradigm was revisited in a study by Pavel Goldstein and colleagues at the University of Haifa, but this time both partners were connected to an EEG in order to determine patterns of brain activation. The theory is that the closer you feel to someone and the more you empathise with them, the more their pain resonates with you. It's tough to measure, but the fascinating idea is that when you're deeply in tune with another person, the patterns of activation in each of your brains tend to match. This is what's known as brain-to-brain coupling and it's been suggested as the basis for mutual sympathy and understanding. Sure enough, when a painful stimulus was applied experimentally to one person, the activity in their partner's brain altered, mirroring what was going on inside their own. Moreover, the closer the coupling of brain activity between the two partners, the more that the sufferer gained pain relief from the simple act of holding their partner's hand.

*

The centrality of touch to our lives is reflected in the words we use. We talk about how we feel or describe ourselves as being 'in touch with our feelings'. When we contact someone, we 'get in touch'. If we're annoyed by someone, we might say that they grate on us, or that they're abrasive. A sensitive person might be said to be thin-skinned, whereas a less socially adroit individual may be callous, or tactless, both metaphors relating to the skin or touch. Then there are phrases such as 'feeling rough' when we're unwell, 'lacking substance' to mean someone, or something, that is unsatisfactorily engaging. We might describe a loving and generous person as 'soft' or 'warm' and their less empathetic equivalent as 'hard' or 'cold'. So much of our language employs concepts relating to tactile sensations, it raises the question of whether our language might be mirroring some profound capability of touch to shape our thoughts.

In 2010, a team of researchers led by Joshua Ackerman of MIT investigated whether our experiences with touch played a role in our judgements. They devised an ingenious series of scenarios, manipulating one tactile aspect in each to see whether they influenced people's perspectives. In the first, they asked unsuspecting volunteers to pass judgement on the CV of a job applicant. The trick was that sometimes the CV was attached to a heavy clipboard and at other times it was presented on a flimsier equivalent. Amazingly, those who read the CV on the hefty clipboard evaluated the candidate as being better overall and more serious. They didn't, however, think that the applicant would necessarily be more likeable, all of which chimes with our use of the word 'weighty' as a metaphor for earnestness and solemnity.

Next, the researchers asked volunteers to read an account of a meeting between two fictional people and to describe their thoughts about the interaction. This time, however, the volunteers were given a task to complete immediately before the reading test. Some of them were given a puzzle whose pieces were coated in rough sandpaper, while others were given smooth pieces. Sure enough, the texture of the puzzle pieces influenced how they described the meeting; those who'd been working with the coarse pieces concluded that the relationship was, like the sandpaper, harsh and difficult.

Ackerman and his colleagues used a similar approach to see whether handling hard or soft materials influenced the perspective of people reading about another interaction. Once again, those who'd held a firm, unyielding object judged the fictional characters to be more rigid and stern than those people who'd been fondling a softer object. A final experimental situation had volunteers engaged in a situation where they were bidding on a car. The first offer they'd made had been rejected and they were asked to make a new offer. The volunteers were tested under two different conditions: in the first, the prospective car buyers were sat on hard chairs, while in the second their chairs were cushioned. It seems that the experience of the padded chair made the bidders themselves more flexible in their behaviour; while both groups of participants increased their offer, those in the comfy chairs made a bid that was on average around 40 per cent higher than that offered by those sat on wooden chairs.

Taken together, these outcomes suggest that we can be subconsciously influenced by tactile experiences. In the same way, when we're holding a warm drink, we feel correspondingly warmer towards those people we meet in social situations than when we're holding a cold drink. In other words, the material metaphors that we use in our day-to-day lives can be seen to emerge directly in our thoughts and behaviour. Touch something rough and it coarsens our attitude, touch something unyielding and our judgements become harder, and so on. It's an extraordinary demonstration of our senses shaping our perspectives.

*

Touch is the domain of our largest and most multifunctional organ, the skin. Our bodies are covered in something like two square metres of the stuff, and it's heavier than you might imagine, comprising around a sixth of our total body weight. It marks the boundary between us and the world beyond. It protects us from an army of pathogens. It insulates us and keeps the precious fluids within from escaping. But skin is much more than simply a covering, or a barrier – it's a giant sense organ. Embedded within its many layers are batteries of diverse sensors, each specialising in a different task. It's these that allow us to feel pressure against the skin, as well as vibrations, itches and tickles. There are many different types of touch, but our singular perception of them emerges from the interplay between four different types of receptors.

Some of the most sensitive aspects of touch stem from oval-shaped receptors just below the surface of the skin, known as Merkel cells. When we brush our fingers against an object or hold it in our hands, these cells are responsible for much of our tactile sensitivity, providing sophisticated feedback on everything that we touch. They're incredibly sensitive, capable of detecting skin displacements, indentations if you prefer, of less than a single micron,* which is 100 times less than the diameter of a human hair. This sensitivity comes in part from the fact that Merkel cells are found closer to the skin's surface than any of the other receptors. They are scattered across the entire body, but they tend to be concentrated in those areas that we rely on most extensively for fine-scale touch. In the tips of our fingers, as many as 100 of these miniature receptors are packed into every single square millimetre of skin. They're also what's known as 'slow-adapting', which means that they provide continuous updates on whatever we're touching. For instance, as your fingers glide across a keyboard, it's the Merkel discs that provide the ability to feel the edges of each key.

---

* A micron, or micrometre, is one thousandth of a millimetre.

This slow-adapting nature of Merkel cells means they provide an ongoing commentary on everything we feel. Sat a little deeper in the skin, Meissner corpuscles also contribute to what's sometimes called gentle touch. A key part of their job is to detect movements against our skin, but they are fast-adapting so alert us only when things change. It would be distracting if you continuously felt the chair that you're sat on, or the clothes on your back. By registering things only when they change, Meissner corpuscles adopt a kind of 'need to know' approach, balancing the need to keep us informed without constantly bombarding us with information. Like Merkel cells, they're concentrated in those areas that we use most extensively for touch, and one critical role that they perform is to allow us close control on our grip. If we hold a glass too lightly and it starts to slip though our fingers, we'll instinctively tighten our grasp (unless, of course, it's not the first glass we've held that day and our responses are a little slow).

Merkel cells and Meissner corpuscles combine to provide exquisite sensitivity of our sense of touch, our fine-scale perception of the tactile world. Each of these receptors has a small receptive field, meaning that they provide a sharp-focussed, high-resolution view of the world that's in contact with the skin. The sense of touch is, however, nothing if not a team effort and there's another receptor that plays an important role. Pacinian corpuscles are chunky, multilayered sensors buried deep within the skin. They respond to deep pressure and high-frequency vibrations; as we move our fingers across a surface, we pick up its texture, roughness and character as vibrations. It's a bit like the way in which the stylus of a record player reads the miniscule ridges on a vinyl disc.

Dynamic touch, as this is known, is much more informative than its alternative, static touch, which is why when we investigate something by touch, we tend to glide our fingers across it. Yet a remarkable recent experiment performed by Cody Carpenter and colleagues from the University of California, San Diego, indicates that both types of touch are more sensitive than we'd previously recognised. In Carpenter's study, untrained volunteers were asked to attempt to differentiate between materials using touch. It turns out that we're incredibly good at this; even when the difference between the materials amounted to a surface coating that was a single molecule in thickness, the volunteers were able to tell them apart. They were especially discerning when allowed to slide their finger across the surface, but even when they were instructed just to tap the surface with their fingertips, they could still discriminate between them.

Last but not least among the quartet of touch receptors is the Ruffini ending, which is perhaps the least well understood. These are primarily focussed on informing the brain about our posture, which they achieve by monitoring the extent to which our skin is stretched. When we reach out to take a glass from a shelf, it's these receptors that register the extension of the arm, and when we close our fingers against that glass to pick it up, they provide us with the necessary sensation of the hand as it curls about the vessel. Throughout the body, Ruffini endings provide us with the sense of what the body is doing, which is essential for coordinating our activities.

While each receptor has its own strengths, it's the synergy between them that's so essential for touch. Take for instance our hands, the jewels in our tactile crown. Many of us might be guilty of taking for granted the incredible precision with which we handle things. It's something that improves with practice during our early years, as anyone who's seen a child try to shuffle a pack of cards will know. By the time we're in our early teens, we've gained the impressive fine motor skills needed to manipulate objects with dextrousness and sophistication. The precision of touch, what's sometimes referred to as tactile intelligence, continues to be an enormous challenge to emulate in the field of robotics. We can build computers that can perform quadrillions of calculations per second, simulate the birth of the universe, and interact with us in text conversations in such a way that we find it difficult to distinguish them from humans. Yet for all the extraordinary advances in artificial intelligence, it's spectacularly difficult to develop a robotic hand that can smoothly pick up a cup of tea without spilling it, crack an egg, or secure a morsel of food using chopsticks – things that humans do without thinking.

Our hands are precision instruments that have no equal in nature. Furthermore, the finesse with which we're capable of handling things sets the scene for one of the most important steps in the evolution of our species: the development of tools. Though there are other animals that similarly use tools, none has such a reliance upon them. Imaging studies show activation of an area in the brain known as the anterior supramarginal gyrus, which seems to be implicated in our ability to link cause and effect in relation to tool use. Yet while the brain is impressive in this respect, it is touch that furnishes us with the dexterity to work these implements. Touch receptors play an indispensable role in translating the feel of things that we hold as we use them. It might be a knife as we slice bread, or a pen as we write; we gain the impression of that object as being an extension of our own hands. It's this remarkable connection that allows us the ability to perform the kinds of intricate tasks that we rely on every day.

*

What we call touch emerges from the combined activities of the diverse receptors within the skin. It's the brain's job to integrate the different inputs and to render the information into a single coherent tactile perspective. Given the vast number of receptors, this is an incredible feat, though some simple experiments that you can try for yourself can uncover some of the brain's shortcuts. Grab three fairly chunky coins and put two of them in the freezer for fifteen minutes or so. Then arrange the three in a line, with the two cold coins at either end. Place the tips of your index and ring fingers on the cold coins and allow yourself to enjoy the cold for a second or two, before lowering your middle finger until it touches the middle coin. Most people will get the same icy feeling in their middle finger that they've gained from the other two fingers, even though the middle finger has no reason to feel cold.

What's happening is that the brain is filling in the gaps in your perception, inventing a feeling based on the likelihood of it being correct. The brain didn't evolve to consider the possibility that someone might be fooling around with coins in freezers; the most probable explanation is that all three fingers are touching something cold. The brain isn't so gullible though. Until it gains some tactile stimulation from the middle finger by touching the coin, there's no particular thermal sensation in that finger. When the middle finger makes contact, the brain registers that there's some input and delivers the illusory feeling of coldness. If you were to use the middle finger of the opposite hand, you wouldn't get the same experience. So the brain is interpolating a feeling that comes from adjacent fingers, but it's smart enough not to do the same with more distant body parts.

Something similar happens with what's known as the 'cutaneous rabbit illusion'. Hold your arm out in front of you and look away. Then ask someone to deliver a series of taps on the inside of your forearm near your wrist in rapid succession and then to follow it up by doing the same further up the arm, perhaps around the elbow. For most people, it will feel like the taps are ascending the arm, like a bunny is hopping along it. While you're at it, you could try the 'tau illusion'. In this trick, a person is tapped on the upper arm and then on the forearm. Do this twice, once with as little time lag as possible between the taps, and once separating them by a second or so. When the two taps have only the briefest interval between them, the person will likely feel that the distance between the two points is shorter than when the taps are separated by a second or two. In both illusions, the brain is subject to a series of biases in the framework it uses to understand the world.

Under normal circumstances, the time between two similar, successive sensations and the distance between them on the body are very closely related, so the brain uses its past experience to estimate one from the other. Messing with this by changing the timing alters our perception of the distance.

Staying with the theme of trickery, cross your index and middle finger until you form a V-shape at their tips. (This is easier if you have normal fingers, instead of the sausages that I'm blessed with.) Then take a marble, or any smallish object, and put it in the cleft between the two fingers. Effectively you're now touching the object with the opposing sides of each of the two fingers, which can't normally happen. Perplexed by this nonsensical situation, the brain will usually deliver the sensation of touching two separate objects, rather than the one. That's what's meant to happen, anyway. It didn't work for me, which might be because of my ludicrous fingers, but there's probably a better explanation why. Since I had my line of sight trained on my fingers as I did this, there's a sensory conflict between touch and vision. In these situations, vision usually wins out.

This is something that the American psychologist, James Gibson, described in the 1930s. In his demonstration he would give a person a ruler while their eyes were shut and ask them to remark on it, based on how it felt. There was nothing unusual about the ruler, so the person might say it was wooden and straight. So far, so good. The trick was that Gibson's volunteers were wearing goggles fitted with lenses that made straight lines appear curved. When he asked them to open their eyes and feel the ruler again, their tactile information fell into step with their vision – the ruler both looked and felt curved. Even though they knew the ruler was straight, they couldn't dissociate the feeling from the visual input.

Integrating information from different senses, or even from different receptors of the same sense, is incredibly demanding for the brain. The use of basic rules to simplify the process of converting sensation to perception means that the brain can be prone to error, which is what these illusions exploit. Perhaps most importantly, however, these tricks demonstrate that our brains aren't passive when it comes to representing the signals that they get from the senses. Touch isn't entirely in the skin, it's in the mind, too.

*

Alongside the fundamentals of touch, the skin collects information on cuts, pinches, temperature and even chemicals such as capsaicin. All of these are detected by receptors in the skin; as a result, they're sometimes bundled together with touch into a broader modality, one that we know as the somatosense, the sensation of the body.

Nestled among all of the specialist Merkel discs, Meissner corpuscles and their like are a host of nerve tendrils, doing a very different job from these touch receptors. These are the so-called free nerve endings. They're free in the sense that unlike the touch receptors, they don't terminate in some micro-anatomical grand gesture. Pacinian corpuscles look like onions, for instance, with their sophisticated layering, while there are no prizes for guessing the shape of a Merkel disc. By contrast, free nerve endings branch and taper off in an unshowy way, much like the roots of a plant. These roots reach beyond the touch receptors to sit close to the surface of the skin where their job is to await the arrival of bad news in the form of unpleasant stimuli.

Free nerve endings act as nociceptors; they detect and relay signals that we might feel as pain. That said, nociception (from the Latin word *nocere*, meaning 'to harm') isn't synonymous with pain. For one thing, we can feel emotional pain that's independent of any bodily damage. For another, even when we are injured, there still remains a difference between nociception and pain. Essentially, nociception is the means by which the body senses and encodes noxious stimuli into nervous messages, while pain is what the brain does with those messages and it's a subjective experience. Nociception is an essential part of the body's defence system, protecting us from harm. To understand the difference between the two, think about a carelessly dropped drawing pin ambushing an unwary foot. If you stand on the pin, a message from the injured foot shoots to the central nervous system and activates a motor reflex so that you jerk your leg upwards and away from the pin. Since it's a reflex and not consciously experienced, most of us assume that we leap upwards because of the pain, but in the milliseconds that it takes us to remove our foot we might feel no pain at all – the pain comes fractionally later. Having these reflexes means that we're far more effective at minimising injury than we would be if we were to rely on the sensation of pain to get clear of the pin.

The brain itself, comprising something like 85 billion neurons, has no ability to feel pain, and yet all pain is created in the brain. Ah, you might say, what about a headache? Well, headaches are triggered by nerves in the muscles and blood vessels in and around the brain, rather than the brain itself. So while the brain provides our experience of pain, the organ itself is immune to it. Nonetheless, our state of mind has a massive influence on the feeling of pain. We each feel pain differently according to our genetics, health and even attitude. Tellingly, the same person might perceive an identical injury occurring at different times as causing more pain, or less. That's because pain isn't solely a physical experience, it's intertwined with psychology so that the way we feel on a given day will influence our perception of it.

Pain is also affected by the catalogue of injuries in our life history. Once as a child I was scampering home when my mind conjured the idea that I was being pursued by some nefarious, fleet-footed, multiheaded creature with an appetite for ragamuffins. I covered the last 100 metres in a panicked blur and, reaching home, I turned the door handle and shot inside. Except I didn't. The door was locked and I ran through the glass part of it, lacerating my knee. It was my first serious injury and my original introduction to deep pain. I can remember the sudden shock of the pain with absolute clarity; a trace of it still lingers whenever I damage that knee, which now that I'm an adult and don't fling myself around after stray footballs or similar, is infrequent. Still, it's there and it's what's known as a pain memory.

The collection of such pain memories that each of us accumulates in our lifetimes plays a role in our anticipation of future pain, as well as our anxiety and experience when pain is current. One of the most amazing aspects of pain memories can be seen in people who've undergone the amputation of a limb. Even though the limb is no longer there, many report that they can still feel it, often to the point of pain. This seems to result from the brain storing this experience of pain in the damaged limb before amputation as a pain memory. Once the limb has been removed, the brain is left with a sensory mismatch that it struggles to resolve. Treating the pain effectively with local anaesthetic before amputation seems to help limit the formation of these pain memories and so reduce subsequent phantom pains.

*

I have a particular pain memory arising from an unfortunate encounter with an angry insect. Early in my career I was involved in a field course, training undergraduate biologists in the study of the hardy creatures that eke out a living in the surrounds of Malham Cove in Yorkshire. The animals who inhabit this beautiful but stark place must be tough to put up with what seemed to me to be near incessant rain and a biting wind. My task one particular day was to use a dipping net to find whatever rugged beast might be lurking in the frigid ponds of the high moor. I managed to catch a few specimens, but the star of the show was a larva of a great diving beetle. These are consummate hunters about the length of your thumb and they're strangers to fear. They don't think twice before tackling animals much larger than they, armed as they are with a pair of fearsome mandibles that curve from either side of the head and taper to sharp points. Having been caught and transferred to a viewing aquarium, the predator was in a terrible mood and sought to take out its frustrations on anything that might cross its path.

I was peering at it when one of the students appeared at my shoulder and asked, 'What would it do if I put my finger in front of it?'

I was slightly nonplussed, eventually replying with what to me seemed obvious: 'It'd bite you.'

'Would it hurt?' he asked, winning the prize for the daftest question of the day.

'Definitely.'

Inexplicably, the student plunged his hand into the tank and put his finger enticingly before the beast, which acted on cue and latched onto the student's digit, drawing an expletive, some blood and a rapid withdrawal of the offending hand. Regrettably, the larva was attached to his finger, meaning that I had to intervene. I gingerly prised open its mandibles, being careful not to injure the creature. Robbed of its victim, however, it sought recompense and bit me.

As with many sudden injuries, the pain came in two waves. A sudden sharp zap of pain accompanied the bite, making me jump. I detached the doughty insect and returned it to its quarters, by which time the acute sensation had given way to a deeper, duller pain. Injuries are often characterised by these distinct waves of pain. The first zing is fired off to the brain by nociceptors along super-fast Type A nerve fibres, carrying the vital information that the body has been harmed. The second travels, like the feeling of emotional touch, along slower C-Tactile fibres and in effect describes the extent of the damage. These latter signals hang around for some time, prompting a cascade of behavioural and physiological responses, including in this instance, my barely masked contempt for the student.

I rubbed furiously at the wound, a common response to such injuries and something that we've likely been doing for as long as our species has existed to ease the immediate pain. The hurt that we experience is the result of local nociceptors sending their distress signal to the brain. By rubbing or holding the affected area, we transmit other, non-pain-related tactile messages, so that the brain is confronted with a plethora of different stimuli and has a lot more to think about than simply the zinging discomfort. Instead of taking centre stage as the solo sensory act, the pain message simply becomes one of many. And while the pain doesn't magically disappear, it does at least temporarily lessen in the melange of nervous input.

All manner of things are capable of causing us pain. Some, like the biting larva, cause mechanical damage. This is one of the three categories of noxious stimuli that our nociceptors are on the lookout for, thermal and chemical being the other two. Mechanical stimuli include everything from dropping a book on your toe to slicing your thumb while inexpertly chopping up vegetables. Thermal nociceptors in the skin start to respond whenever temperatures at skin level depart from a narrow band of comfort, between 20°C and 43°C.

The further we get from this range, the more the nociceptors signal their displeasure. For instance, climbing into a bath that's 55°C is an intense experience and we'll feel pain in less than a second. A 45°C bath would still be uncomfortable, but it takes something like seven seconds to experience discomfort. Feelings of heat and cold are generated separately by two different receptors. Cold receptors, with the catchy name TRPM8, aren't only excited by temperatures below 20°C; menthol sets them off too, which is why people often describe mint as tasting 'cool'. The skin has to drop to around 15°C in most people before we start to experience pain. In a similar way, the heat receptors, chief of which is TRPV1, aren't only triggered by high temperatures but also by fiery chillies.

Chemical nociceptors in the skin respond to harmful sub-stances, which includes substances released by our own bodies following tissue damage. When you cut or bruise yourself, chemicals such as arachidonic acid are released by the cells in the area, in turn stimulating nociceptors and making the wound site sensitive to pain. It might seem harsh that our own body conspires against us to make injuries feel more tender, but it's a vital means of activating behavioural change. Our extra caution when we're nursing an injury helps prevent further damage at the site, allowing time to heal. Nonetheless, it's unpleasant, and we've countered nature by devising painkillers like ibuprofen that work by blocking the chemical pathways that the body uses to signal pain and damage.

Alternatively, we might encounter noxious chemicals in the form of a stinging insect or things that we willingly put in our mouths, such as chillies, mustard or wasabi. Typically, these chemicals are produced by plants to defend themselves against questing herbivores, and for the most part they're successful in keeping grazers at bay. But the plants hadn't reckoned on a species that incorporates small amounts of different ingredients to make a complex dish. So while few people would derive enjoyment from wolfing down large amounts of Scotch bonnets or English mustard, by adding a pinch of this and a soupcon of that, we get a pleasant tingle from these potent ingredients.

In addition to the caresses of others and the attention we give to our injuries, the body has its own way of dealing with trauma. For instance, many people enjoy the hit of spicy dishes, when the active ingredient of chillies, capsaicin, rolls across the tongue like a bush fire. Part of the reason that it can be so pleasurable is that capsaicin stimulates nociceptors in the mouth, which register the ingredient as a noxious chemical, causing sensations of pain. The brain's response is to release endorphins and, subsequently, dopamine, which provide a feelgood buzz and numb the pain, to some extent.

At the same time, in the heat of the moment, adrenaline has

long been known to give us the temporary ability to withstand the most excruciating injuries. In April 2019, Kurt Kaiser, a Nebraska farmer, slipped and caught his leg in a churning grain auger, a machine that draws grain in through metal blades. Seeing his leg being pulverised by the machinery, with no way to shut it down and no one to call for help, Kaiser knew he was in deep trouble. Then he remembered the pocketknife he carried with him and realised that there was a way out – he needed to amputate his own leg. Determinedly, he set to work, cutting below his knee until eventually, his leg gave way and he was able to escape the auger's clutches. Crawling across rough ground to a nearby building, he was finally able to call for help.

Stories such as these make the news on a reasonably regular basis, and one thing they all have in common is that the victims seldom describe any pain. Part of this is due to the endorphins that I've already described, but the more important role seems to be played by adrenaline. This hormone is produced rapidly in response to dangerous situations, priming the body and putting it into survival mode. Adrenaline isn't a painkiller. Instead, it buys us a short window of time during which our attention is focussed entirely on an immediate threat. It allows us to disregard the pain and concentrate solely on resolving the situation. Once the concentration of adrenaline in the body begins to ebb, the perception of pain returns in force.

*

The skin's sensitivity is a result of constellations of receptors, each served by a vast network of nerves. The dermis is interwoven with hundreds of thousands of individual nerve fibres, each an electrical conduit that carries to the brain messages of every contact we make with the outside world. Not all parts of the skin are equally sensitive, however. Some areas, such as the back, get by with only a skeleton crew of nerves, while others, like the lips and fingertips, have a nervous thread count higher than the most expensive linen. This imbalance is reflected in our brains, and more specifically, in the somatosensory cortex, which devotes itself to areas like the hands, face, and genitals, allocating vast chunks of its real estate to them, while leaving the rest of the body to get by with comparatively meagre neural resources.

The discovery of how the body is depicted in the brain represents one of the great advances in neurology in the last 100 years. The man behind it was a neurological explorer and mapmaker named Wilder Penfield. Unlike those who travelled across continents to chart unknown lands, Penfield's domain was the human brain.

Perhaps unusually for such an accomplished scientist of his times, his early life was emotionally and financially turbulent. His father, a family doctor in Washington State, was a loner, preferring to spend time hunting in the wilderness rather than attend to his patients. The inevitable result was that his medical practice ran into difficulties. Facing hardship, his mother took the eight-year-old Wilder to live with her parents in Wisconsin, and the straitened circumstances that they experienced seem to have been instrumental in shaping his character.

Wilder enrolled at Princeton, where he drifted through his first two years, failing to find inspiration in his chosen subject of philosophy. It was toward the end of that second year that his life changed. Although his memories of his father had initially sworn him away from medicine, he found his enthusiasm kindled by the teaching of the great biologist Edwin Conklin. Penfield had found his calling.

In 1913, by now aged twenty-two and having graduated, Penfield learned that he had failed to gain the prestigious Rhodes Scholarship that would allow him to study medicine. Undaunted, he put in for it again, and this time he was successful. However, this being the age of Classicism in academia, in order to gain entry to Oxford, he had to sit an entrance exam in Greek. He threw himself into the task, trying to master an entirely unfamiliar language from scratch, but once again fell short.

A glimmer of hope remained – he was allowed a resit. In preparation, Penfield showed up to the pathology laboratories of Harvard early each morning to spend an hour being schooled in Greek while surrounded by human cadavers awaiting dissection. His hard work paid off and, claiming his scholarship, he left for Oxford where he was taken under the wing of Sir Charles Sherrington, who would later win the Nobel Prize for his work on the nervous system. It was Sherrington who made Penfield realise that 'the nervous system was the great unexplored field – the undiscovered country in which the mystery of the mind of man might someday be explained!'

Penfield's contribution to the understanding of the brain is vast, but what he's best known for is his mapping of that organ. As he operated on patients, he kept meticulous notes of the parts of the body that responded to his stimulation of the brain. Gradually, he was drawing a map of which location of the brain serves each given part of the body. The idea was not new, but Penfield was the person who brought it to life. One of the most obvious early results was that the parts of the body are not given equal treatment by the brain. Using Penfield's data, a model could be constructed to show what a person would look like if their bodily features directly corresponded in size to the area of brain given over to them. The result is the cortical homunculus, a grotesque creature with a huge head, dominated by an oversized, lascivious mouth, and a pair of truly colossal hands. Penfield himself wasn't a fan, and is supposed to have remarked, 'I would kill the damn thing if I could', yet as an immediately intuitive depiction of our senses, it's hard to beat. It's this model that shows clearly how we sense the world with touch, and how our hands and our lips are such exquisite gauges of the physical world around us, packed as they are with vast numbers of nerve endings.

Notwithstanding his dislike of the homunculus that he created,

Penfield's work has provided arguably the most accessible depiction of the brain's workings that is currently available in relation to any of our senses. He died in 1976, leaving an incredible legacy. Alongside his work on cortical representation in the brain, his contributions to the treatment of epilepsy and brain injuries, as well as his work on visualising the brain, are enormous. His humility was paramount throughout, and he was always at pains to share the credit for his work with those around him, always using the word 'we' rather than 'I' when describing his discoveries. His autobiography *No Man Alone*, published posthumously in 1977, reflects the team ethic that this great man always emphasised.

While Penfield's work centred on examining the brain to understand how different parts of the body were represented, if we're to understand touch fully, we need to see how tactile receptors are distributed about the body. This is no easy task – it's one thing to determine how many sensory receptors there are concentrated and clustered in the retina or the inner ear, but quite another to tackle something so comparatively vast as the skin. Nonetheless, painstaking work has been done to examine just how richly supplied with tactile nerves each part of the body is. The nerve fibres that reach up to the skin are minute, some less than a thousandth of a millimetre in diameter. You could bundle 400 of them together and they'd still be less substantial than a single human hair. This is nervous wiring at its finest, which is why we don't typically see nerves when we cut ourselves. Despite their insubstantial nature, these nerve fibres carry messages from tactile receptors in the skin to the brain. The more nerves in a given area, the more receptors they connect to and so the more sensitive that area of skin will be.

A recent estimate by Giulia Corniani and Hannes Saal of the Active Touch Laboratory at the University of Sheffield puts the average number of sensory nerve fibres that serve each single square centimetre of the human body at around fifteen and the total for the whole body at around 230,000. But as Wilder Penfield discovered in his explorations of the brain, they're not equally distributed. The fingertips, for instance, squeeze 240 fibres into every square centimetre, while the torso gets by with a measly nine in the same space. In fact, nowhere on our bodies gets anywhere near the sensory richness of the hand. (I know what you're thinking, and the answer is no, not even those bits.) The next closest to the hand in terms of neural richness is the face, especially the area near the lips, which have around eighty-four fibres per square centimetre.

These nerve-fibre counts in each area of skin match up reasonably well to the size of that area's representation in the brain, but the relationship between the two is far from perfect. Amazingly, even with the throng of nerve fibres that permeate the hands and face, for the brain, this still isn't enough. It further magnifies the input from these zones and pays them extra attention, doting on them in the manner of a teacher attending to their favourite pupil.

That said, the brain's favours can shift, and this is at least in part dependent on usage. The more we use some part of the body in tactile tasks, the more the brain reorganises to upgrade its representation of that area. Players of string instruments develop extraordinary dexterity with the fingers of their left hands, which modify and control the notes as they play. Examining the brains of violinists shows that the representation of these digits in the brain is substantially greater than for non-musicians and the longer they've been playing, the greater this discrepancy becomes. Similar patterns can be seen in people who make extensive use of touchscreens, and in particular, mobile phones. The more they use their phone, the more the brain upgrades its focus on certain digits, especially the thumb. The generation that grew up with iPhones and the like have shaped their brains to fit their world.

While there is a degree of flexibility, it remains the case that certain areas are specialised for touch. The dense tangle of nerve fibres that can be found at the fingertips serve an army of tightly packed tactile receptors. This means that we're much better at discerning fine detail with the fingertips, since each separate receptor reports on a very small area of skin. In other parts of the body, where the fibres are more spread out, each receptor has a much greater area to cover, making our touch resolution much less sharp. One common measure of this is known as the two-point discrimination test. The idea behind it is simple; a participant is gently prodded simultaneously with two wooden toothpicks, or some more scientifically defensible pointy objects. When the two sticks are separated by a reasonable distance when they prick the skin, it's easy to tell that there are two separate jabs. As they get closer together, however, it becomes increasingly difficult to tell whether you've been prodded once or twice. In areas like the back and the thigh, with any distance less than about 4 centimetres between the two points, it will be hard for most people to tell if they got two jabs or one. On the cheek or nose, the distance drops to a little under 1 centimetre, while at the fingertips, many can distinguish two jabs even if they're only a couple of millimetres apart.

The free nerve endings that do the job of nociception are the most common receptors in the skin, vastly outnumbering those that provide our tactile awareness of the outside world. It's often been said that those parts of the body that are most sensitive to touch are also most sensitive to pain, and there is some truth to this. The palm of the hand and the insides of the fingers provide an incredibly sophisticated sense of touch, but they're also correspondingly highly vulnerable to pain. Presumably that's why in bygone days of corporal punishment in schools, sadistic teachers would often cane errant children on their hands. There's a similar close relationship between touch and pain on the soles of the feet, again a favourite of torturers, as well as on the face.

Generally, our smooth, glabrous skin is replete with nerve fibres that provide exquisite sensitivity for both touch and nociception. Elsewhere in the body, the picture isn't so clear. Tactile acuity tends to be focussed on the parts of the body most often used for touching; the further you go along the arm towards the hand, the more innervated the skin becomes. In contrast, nociceptors are found at greater densities nearer the torso; while touch becomes more acute as you descend the arm, for pain it's the other way around. Exactly why this should be isn't known, though it could be simply because, using the same principle as a bulletproof vest, nociceptors concentrate on protecting the body's most essential areas.

The estimate of almost a quarter of a million microscopic nerve fibres threading though the skin doesn't apply to all of us, sadly. In our early years, the number of nerve fibres proliferate as we grow, peaking in our teens and early twenties. After that, time's familiar sensory slide begins. We lose something like 8 per cent of these nerve fibres with each passing decade; by the time we're eighty, the number has decreased by around a half. Worse still, the losses are most dramatic in our hands, face and feet, the very places that provide such exquisite sensation when we're young. At the same time, Meissner corpuscles and Merkel cells are lost from the fingertips at an alarming rate, meaning that the sensitivity of our sense of touch declines as we age. We might take some solace from the knowledge that we can use the brain's extraordinary adaptability to train our sense of touch, but the fact remains that there are substantial age-related differences between us. These, as we'll see, are not the only differences.

*

One of the most remarkable things about the senses is the extent to which we all differ in our capacity to perceive the world. Blind people, for instance, have long been known to display remarkable acuity in their senses of smell and hearing, and there's plenty of evidence to support the idea that they have a faster and more accurate sense of touch, too. This super-sensitivity to touch arises partly because of the brain's plasticity, adapting to augment sensory inputs other than sight. People who have been blind from birth often have greater tactile abilities than those who lost their sight later in life. The sensory brain is reconfigured to provide greater support to senses other than vision, and this advances in harness with the extra perceptual effort being channelled through the sense of touch. For instance, those who read Braille often display greater sensitivity in the fingertips of the hand that they favour while reading. As Helen Keller put it, 'Touch brings the blind many sweet certainties, which our more fortunate fellows miss because their sense of touch is uncultivated.'

Just as with the other senses, women generally have a more sophisticated ability to discriminate by touch, though this doesn't seem to result from any fine-tuning of their sensory sophistication. The reason for the disparity is the basic fact that, on average, men tend to be larger than women and so have bigger hands. Indeed, if you were to pick any two adults of the same age and then compare their tactile acuity, the likelihood is that the one with smaller hands would do better. That's because the sensitivity of touch depends pretty much directly on the concentration of nerve endings. Since each of us have roughly the same number of nerve endings but a different amount of skin, the more petite of us have a higher density of nerve endings and so gain the benefit of a more finely tuned sense of touch. Put more neatly, and alliteratively, in the title of a paper by Ryan Peters and colleagues in 2009, 'diminutive digits discern delicate details'.

So that's touch, but are there any differences between the sexes in regards to pain? This, of course, is a topic on which many have opinions, and which can descend into a fatuous discussion of who's tougher: men or women? It's left to researchers to examine the question in an objective manner and to collect data on pain thresholds. Many such comparisons have been run, with the findings published in periodicals with wonderfully stark titles such as *Pain* and *The Journal of Pain*. They describe how, over the years, volunteers, presumably of a masochistic disposition, have lined up to be subjected to all manner of attacks in the name of science, from being subjected to blunt force trauma and being jabbed with sharp things to being given electric shocks, being burned or frozen, and having chilli juice rubbed on delicate skin. The results are clear and consistent: women are significantly more likely to display a sensitivity to pain.

This aligns with other studies reporting that women are more likely to take painkillers, that they suffer more conditions such as migraines and chronic pain, and that they're more likely than men to go to the doctor to discuss their pain. Does this mean that women are innately delicate, fragile and weak creatures? No, it absolutely doesn't, and here's why. Pain, as mentioned earlier, is a subjective experience. Studies of pain rely on people describing how much something hurts on a scale of 'shrug' to 'scream'. This doesn't invalidate the research, but it does make a fair comparison harder. We're all products of the societies in which we live, and it remains true that expressing pain is generally less socially acceptable for men than it is for women.

Moreover, the experience of pain is influenced by hormones. Exposure to testosterone during early development tweaks a range of different nociceptive pathways, ultimately making males less sensitive to pain. Women, however, receive far less of a boost from sex hormones such as oestrogen and progesterone. There's evidence that when oestrogen concentrations in the blood are high and those of progesterone are low, as is the case immediately before ovulation, women's pain sensitivity is reduced. At other times of the menstrual cycle the picture is far more complex, so that the same stimulus could end up producing a different experience of pain on different days. What's certain though is that women are far less insulated against pain by their hormones. It's not a matter of being weak; pain is a different experience for them.

Things for redheads are even more complicated. Their red hair and pale skin emerges because of their comparative lack of the pigment melanin.* But the biochemical pathways that produce melanin also have an unexpected effect on pain thresholds, leaving red-haired people with a higher resistance to pain. Strangely enough, they're also less susceptible to some anaesthetics, which means that while most of us are comfortably numb in the dentist's chair, redheads given the same painkilling dose may end up feeling distinctly vexed by the drill.

Aside from gender differences, and those that depend on hair colour, there are innumerable other disparities between us when it comes to our experience of pain. For instance, we know that those suffering with anxiety and depression are hit with the double whammy of being more sensitive to pain. Those who smoke or are overweight also have lower pain thresholds. Meanwhile athletes and people who do intensive, regular exercise benefit not only from higher levels of fitness, they're also able to withstand pain better. In the end, one person's experience of pain is distinct from another's and it'd be worthwhile remembering that whenever we're tempted into judging someone else's suffering.

*

Sometimes pain can be alleviated by the touch of another. In April 1987, when Princess Diana visited the UK's first dedicated HIV/AIDS ward, the country was in the grip of paranoia about the disease. Though it had been determined some years before that AIDS couldn't be transmitted through casual contact, the public had been whipped into a frenzy by scaremongering tabloid headlines and misleading reports that screamed the direst warnings about what was termed the 'gay plague'. The stigma of the disease was such that the patients went into hiding as the TV cameras followed the princess into the ward. Eventually one of them, a

---

* Or more specifically, eumelanin. Redheads do possess a different kind, known as pheomelanin.

thirty-two-year-old man in the later stages of the illness, volunteered to meet the princess and in the subsequent photos, he was shown greeting her with his back to the camera. But what caught the imagination of the press and public alike was the fact that the princess shook the man's hand without wearing gloves. It seems extraordinary that this was an important detail, yet at the time it was considered a brave and even radical move. Many on the medical front line at the time described it as being a turning point in the public perception of AIDS, the moment when a simple, human touch brought sufferers of the disease in from the cold.

Over three decades later, we have been facing a new and different epidemic and it's another in which touch plays a critical role. Social distancing during the Covid-19 pandemic has limited our ability to hug and embrace those beyond our immediate family, but social touch has in fact been gradually seeping out of our lives for years. Some of this is the result, it's claimed, of concerns about being seen to act inappropriately. Doctors have been warned against hugging their patients and teachers hesitate to touch pupils in case their motives are misconstrued. At the same time, younger generations interact more and more with screens rather than with each other, sealing themselves off from the world.

The screens themselves are, of course, an incredible piece of touch technology, but it's a passive, one-way experience. Will the day dawn when we can touch our loved ones remotely? This tantalising prospect has engaged the attention of researchers for a few decades now. One of the first efforts in this direction was the extraordinary Telephonic Arm-Wrestling System in the mid-80s. The idea was that it would allow a pair of determined and avid competitors to challenge one another while being thousands of miles apart. Each of the isolated wrestlers was equipped with a lever as a stand-in for their distant adversary's arm. The force with which they grappled with their lever would be translated into their opponent's lever via information fed through the phone lines. Astonishingly, it didn't catch on but now the technology has advanced to the point where we might actually be on the cusp of an interactive touch experience.

With the arrival of 5G and the ultra-high-speed communication that it allows, the so-called Tactile Internet has moved from the realms of science fiction to reality. Human touch will soon be able to operate without boundaries. We'll finally be able to use the dextrousness of touch to manipulate objects remotely. More than this, we'll be able to feel what it is that we're touching. The possibilities of the Tactile Internet range from a new sensory frontier in internet shopping, where we might feel the texture of materials, to cases where doctors might examine inaccessible patients. There's even the potential to use the technology to convey a caress to a loved one. The downside is that while such a touch would be instigated by a person, it would be delivered by a piece of hardware – an android hand, for instance. Better than nothing, certainly, but still one step removed from the real thing.

Touch has long been, and remains, our most intimate means of connecting with others, but the extent to which we do this depends to some degree on the culture we're brought up in. In the 1960s, the Canadian psychologist Sidney Jourard studied how people interacted as they sat in coffee houses around the world. As Jourard watched, he recorded the number of times that people touched each other during their conversations. His results showed some major differences between us. People in London were the most reserved – they didn't touch each other at all. Those in the US were a little warmer, touching on average around twice per hour, while in France and Puerto Rico people could barely get their hands off each other, touching 110 and 180 times per hour respectively.

Fifty years on, another study revisited the question and found that in the US, people were now touching each other nine times an hour. Interestingly, there was a difference between towns and cities, with the urban dwellers being more tactile than their country cousins. Even so, nine times an hour isn't much and when you consider the huge drop-off in social contact prompted by the Covid crisis, it's little wonder that so many people report feeling chronically deprived of touch. In recent times, the term 'touch starvation' has been coined to describe how the loss of touch from people's lives leaves them feeling lonely and adrift, in turn leading to psychological trauma.

Being starved of touch places us at risk of missing out on a vital part of the social existence that supports us. Being touched leads to an improved immune system, and the release of serotonin and oxytocin, which calm us and make us feel good. It's one of the simplest, yet most profound, things that we have, the most effective salve for the pain of social exclusion and the bedrock of our relationships.

Our skin both separates us from other people and acts as the interface for one of our most essential means of communication. While we can stifle our other senses with blindfolds, or ear plugs, touch is always there to remind us of the outside world in all of its character and texture. More than any other, it's the modality that provides us with a sense of our own distinctness, of our discrete individual identity. Touch is our most profound sense, and it's one that we neglect at our peril.

# The Kitchen Drawer of the Senses

*Magic is really only the utilization of the entire spectrum of the senses. Humans have cut themselves off from their senses.*

– *Michael Scott,* The Alchemyst

So there we are. Vision, hearing, smell, taste and touch. The big five senses that each of us instinctively understands. It would, perhaps, be neat to draw a line there. But although this approach has the advantage of clarity, it neglects the many other senses that we use to navigate our lives. Where does such a view leave our sense of balance, for instance? Is this a semi-sense, or worse, a non-sense? It does at least have a system of receptors to facilitate it. Then what about our sense of time, our ability to work out how long it's been since we started boiling an egg? There's no sensory apparatus here, rather it's just a facility in the brain.

There are a large number of candidates that might claim membership of the sensory club. The picture becomes even more complicated when you realise that some authorities subdivide senses such as touch into many discrete senses. As a consequence, the answers that you might get when you ask how many senses humans have ranges from the basic five all the way up to fifty or more.

If you then pose the question 'How many senses are there?', that number grows further still. These are senses that have evolved in animals other than ourselves, providing insight into challenges that we are largely ignorant of. An exploration of these unfamiliar senses, along with those that we possess yet underutilise, gives us a perspective far greater than that which a rigid adherence to the five-sense orthodoxy ever could.

*

Animals are sometimes described as having 'supersenses', and in many instances these relate to natural phenomena. On the morning of 26 December 2004, a huge rupture occurred at the fault along two continental plates between the Indonesian islands of Simeulue and Sumatra. The energy released was, by some estimates, over 20,000 times greater than that of the bomb that devastated Hiroshima, and it infamously generated a tsunami that caused destruction throughout the Indian Ocean. As it thundered through Aceh, the wave reached 30 metres in height, equivalent to a nine- or ten-storey building. Across the entire region, coastal towns were destroyed by a relentless surge of water and debris that claimed the lives of almost a quarter of a million people.

In the weeks and months that followed the tragedy, one question kept recurring: why had there been no warning? Though Aceh had virtually no time to evacuate, people in places further afield might have been saved had the alarm been raised. It was an hour and a half before the tsunami came ashore in Thailand, and two hours until it hit Sri Lanka. The element of surprise meant that fatalities were far greater than might have been the case. There were no warning systems in the Indian Ocean at the time and while new technology has now been deployed in the region, tsunamis remain notoriously difficult to detect at sea. In deep water, this most deadly tsunami in history was no more than a hump of water, less than a metre in height as it rolled toward unsuspecting populations in the region.

A UN report published in the aftermath of another devastating tsunami that hit the Indonesian island of Sulawesi in 2018 urged against an overreliance on technology. The authors' caution was based on the inaccuracy of systems that log the size of tsunamis out at sea, as well as the difficulties in relaying information across large stretches of at-risk territories. At our current state of knowledge, the many different variables that combine to determine the probability and extent of risk makes accurate predictions an enormous challenge. There is, however, a simpler solution that deserves consideration, at least as an adjunct to our current methods.

Long before a tsunami strikes, animals seem to be aware of the danger. Eyewitnesses of past disasters have described panicked cows and goats charging towards higher ground well in advance of a surge, and flocks of birds departing trees fringing the ocean. It has often seemed as if they are reacting to some stimulus that we're unaware of, one that precedes the arrival of the flood by at least several minutes. If they're sufficiently attuned to the behaviour of animals, local people might take heed and follow them to to safety.

As a case in point, the island of Simeulue was close to the epicentre of the 2004 earthquake, yet among a population of some 80,000 people, only seven died in the tsunami, an outcome that owes much to the attentiveness of the inhabitants to the behaviour of the local fauna. The animals could feel the tremors of the earthquake and may also have been able to detect some other signal, perhaps the infrasound produced by the seismic disturbances that foreshadow earthquakes. Tsunamis also generate infrasound, alerting those creatures able to perceive these deep sound waves to the imminent danger of a deadly wave of water.

History is littered with accounts of animals acting strangely in advance of natural disasters. In the days leading up to an earthquake in the northern Chinese city of Haicheng in the winter of 1975, cats and livestock began to behave unusually. Most perplexing of all, snakes emerged from underground hibernation, only to freeze to death in their thousands. More recently, an entire population of toads who'd gathered at Lake San Ruffino in Italy to celebrate spring in the time-honoured way by the enthusiastic begetting of tadpoles left the water en masse in the middle of breeding. Five days later, a huge earthquake tore through the area. Their sensitivity to seismic shudderings may have pre-warned the toads, though other changes occur in advance of earthquakes, such as the release of gases and electrical energy that results from the grinding and splitting of rocks during tectonic activity. At other times and places, rats have emerged onto streets in daylight, birds have sung at the wrong time of day, horses have stampeded, and cats have moved litters of kittens. In some cultures, especially in areas that regularly suffer such events, these kinds of observations have been incorporated into folklore, enabling traditional knowledge to protect the local populace.

Can technology build on this, using the sensitivity of certain animals to the subtle signals of impending danger? Martin Wikelski, director of the Max Planck Institute of Animal Behavior in Konstanz, believes it can. Over the course of his career, he has developed an extraordinarily sophisticated system that traces the movement of different species around the globe. Each individual animal carries a state-of-the-art tag that transmits detailed information, including speed, acceleration, activity and location. This information is collected by sophisticated aerials on the International Space Station and relayed back to Earth. One of the main goals of the project, known as Icarus, is to study long-distance migrations, and to examine how animals interact with the ecology of their environment and with each other, ultimately allowing targeted conservation efforts. The unprecedented richness and quality of the information, however, provides a means to harness animal behaviour as an early warning system for natural disasters, or, to give it the name that Martin coined, Disaster Alert Mediation using Nature (DAMN).

Some years ago, Martin and his colleagues travelled to Sicily to confront the island's perennially troublesome volcano, Mount

Etna. On the flanks of the volcano, goats graze contentedly on the vegetation that flourishes in the rich, volcanic soil. To mine this caprine local knowledge, a handful of these animals were fitted with electronic tags, allowing the researchers to monitor their behaviour from afar. Martin and his team didn't have to wait for long, as Etna erupted a few weeks later. Retracing the behaviour of the goats in the run-up to the eruption, Martin identified a clear response around six hours earlier, when they became unusually active.

As a scientific measure, however, 'unusually active' doesn't really cut the mustard. So the next step was to establish the exact behavioural parameters that would indicate that the goats had sensed that Mount Etna was about to erupt. If this were achieved, the goat-powered alarm system could then be automated, triggering an alert whenever specific aspects of the animals' behaviour surpassed a threshold value. Over the next two years, the doughty goats successfully detected almost thirty volcanic stirrings, seven of which posed a significant danger. That on its own is impressive, but more was to come. Etna is ringed with measuring stations that use mechanised sensors to predict volcanic activity, yet the goats outperformed these by sensing Etna's disquiet far earlier than the tech gizmos. What's more, they were able to identify the likely severity of the imminent eruption, something that has been notoriously difficult to achieve via scientific instruments. By melding cutting-edge technology with the evolved 'supersenses' of animals, Martin has brought a rigorous twenty-first-century perspective to long-established cultural lore, one which promises to provide an inexpensive and effective solution to a global problem.

*

Martin isn't the first person to try to harness the sensory sensitivities of animals. The ability of animals like leeches to detect barometric changes was put to good use by the aptly named Victorian inventor, George Merryweather. His forecasting device, known as the Tempest Prognosticator, was an attempt to predict stormy weather. The Prognosticator looked like a miniature merry-go-round with small bottles instead of wooden horses, each of which was home to a single leech. In their natural habitat, these creatures remain in damp refuges awaiting wet weather; the arrival of rain piques their interest, while a thunderstorm positively thrills them. In Merryweather's contraption, a low-pressure front would titillate the leeches and they'd respond by climbing the walls of their glass bottle. This had the effect of unbalancing the mechanism, making it rock and ring a bell, alerting nearby people. It's an ingenious idea, and its inventor lobbied hard, though unsuccessfully, for the device to be adopted by the coastguard. He was, however, accorded the honour of seeing it exhibited at the Great Exhibition of 1851.

Leeches are only one of many animals that are sensitive to changes in the weather. In parts of France, farmers once kept frogs in glass jars, exploiting the amphibians' tendency to let out a series of croaks in anticipation of approaching rain. Plenty of plants are similarly good at forecasting the weather. A clover cautiously folds in its leaves as a protective measure ahead of a shower, while marigolds, sunflowers and scarlet pimpernel close up their flowers, possibly as a way of keeping their pollen dry. It's long been known that we, too, are susceptible to feeling under the weather, literally, when the barometer falls. In particular, the intensity of migraines, rheumatism and chronic pain seems to coincide with low pressure events. It's not clear why this happens, though it does seem that a drop in pressure causes the body to release stress hormones and to ramp up nervous activity, in turn making us more sensitive to pain. Whether this amounts to a sense in the strictest terms is questionable, in part because there seem to be no receptors specifically dedicated to it. That said, we know that in mice at least, changes in pressure trigger nervous activity in the inner ear, and something similar may be true in humans.

Storms not only accompany fluctuations in barometric pressure, they also generate electrical disturbances. Bees are alert to this, and the approach of a thunderstorm sends them hightailing back to the hive. The ability to detect electrical fields, known as electroreception, also forms part of a bee's foraging strategy. In common with other insects, as they beat their wings their bodies build up a slight buzz of static charge. When they land on a flower, the charge is transmitted to the plant where it remains before slowly dissipating through the stem and into the earth. Other bees casting around for nectar can detect the electrical activity left by a previous forager and might sensibly reject a charged plant because the earlier visitor will likely have collected much of its nectar.

Strictly speaking, the ability of bees to detect electricity isn't electroreception, since they perceive it by means of their sense of touch, but many other animals do have sensors dedicated to identifying electrical fields. Sharks are famously adept at this, surveying the pulses and hums of charge that surround them as they cruise around their environment. A small fish might hide itself from view, but it can't prevent itself from broadcasting tiny voltages of nervous energy, surrendering its position to the shark. Similarly, mammals such as dolphins and the egg-laying monotremes are capable of homing in on the tell-tale electrical fields of live prey. The wonderful duck-billed platypus is the master predator of Australia's creeks and billabongs. The turbid water sometimes found in these places represents no obstacle to the platypus, whose bill is studded with receptors allowing it to locate tasty invertebrates even when aquatic visibility rivals that of gravy.

Electroreception is a sense that we lack, but we're not entirely unaware of electromagnetic fields (EMFs). The question of just how susceptible we are to them has given rise to all manner of conspiracy theories. Most recently it was claimed that there was a link between the roll-out of 5G networks and the appearance of

Covid-19. There is no link, of course, nor any credible science that supports a connection. Nevertheless, some rather more balanced concerns about the effects of EMFs on humans have been the subject of one of the most intensely researched fields in science. Since the beginning of the twentieth century, power lines, indoor lighting and home appliances have brought us into close contact with radiating energy in the form of electrical fields.

Is this risky? Lots of attention has been devoted to this, testing things like wireless devices and mobile phones and the consensus among scientists is pretty clear – if there is any risk at all, it's vanishingly small, especially when viewed against the day-to-day perils of exposure to sunlight, sustained heavy drinking, or a poor diet.

Still, that doesn't mean we should be cavalier about it. Whether you decide to swathe yourself in tinfoil, wear amulets with questionable properties, surround yourself with crystals, or just live with it is very much a personal choice. The dangers of EMFs seem all the more menacing for the fact that we're exposed to them more or less involuntarily, and we can't sense them. There is however one kind of EMF that animals, and according to some, us, are able to sense. It extends across the entire planet and we've used it to navigate for centuries: the geomagnetic field.

*

In 2013, a team of Czech researchers published the results of a painstakingly collected, and rather unusual, study on the lavatorial habits of dogs. In total, they watched nearly 2,000 toileting terriers, bog-busy bulldogs and Labradors-on-the-loo, before releasing their results to a bemused public. A couple of years later, I found myself in an auditorium in Prague, listening to the research group's leader, Hynek Burda – a man who looks more like Santa Claus than St Nick himself – describe the prize they'd won in the aftermath of their publication. While it's always nice to gain recognition, the Ig Nobel Prize is satirical. It's awarded by the Annals of Improbable Research for 'honouring achievements that first make people laugh, and then make them think'.

Not that this troubled Burda; he was delighted with the scientific infamy and regaled his audience with tales from the stranger side of science for over an hour. The dog study was focussed on magnetoception, the response of animals to the Earth's geomagnetic field. Oddly, Burda and his team had found that dogs line themselves up along a north–south axis before they defecate. It doesn't matter to them if their head is north and their bum is south, or vice versa, but apparently they like to keep things in line. Dogs are by no means the only animals to guide their behaviour according to an inbuilt compass. Burda described how cattle and deer align themselves on the same north–south axis as they graze, and how hunting foxes prefer to pounce in a northerly direction. Like most behavioural data, these animal orientations are noisy. Not every dog that feels the call of nature, nor every chewing cow, is as infallible as a compass needle, yet the patterns are far from random and can't be explained by wind or weather. Perhaps the most telling factor is what happens when animals are located near electricity pylons, which disrupt the geomagnetic field: their orientational orderliness disappears.

Organisms from bacteria to bats have the ability to detect the Earth's polarity. And while the animals already mentioned seem to respond to it in peculiar ways, there are some species, such as eels, pigeons and whales, whose perception of geomagnetism is an exquisitely honed component of their inbuilt navigation systems. But how exactly do animals sense directions? Ground-breaking research on migratory birds suggests that a protein known as cryptochrome-4, found in the avian retina, enables them to see the Earth's magnetic field. That it's a visual experience seems clear; they don't seem able to orient in the complete absence of light, and activation of cryptochrome stimulates the same parts of the brain as regular vision. However, what exactly it looks like to them is anyone's guess. It's plausible that they might perceive it in the form of additional contrast, or brightness, allowing them, literally, to see where true north lies. This extraordinary plug-in to their visual sense makes navigating while flying long distance child's play.

Cryptochrome isn't exclusive to birds. It's an ancient protein that can be found throughout the tree of life. Indeed, it was first isolated in cress, and it's also found in a huge range of animal species, including everything from insects to us. Its primary role in most species is setting the internal clock, through the detection of blue light. In some of these species, however, it has diversified and joined forces with other proteins to form what amounts to a biocompass. But just possessing cryptochromes doesn't mean that animals can flawlessly find their way around. Birds that migrate long distances have cryptochrome compasses that are highly responsive to magnetic fields. By contrast, chickens and homing pigeons seem to have the necessary cryptochromes, but they're just less sensitive than those of other birds. Given pigeons' famed pathfinding abilities, is there another mechanism that allows them to trace magnetic fields?

Iron is a crucial element in the bodies of pigeons, as indeed it is in those of most other animals. It's vital to the haemoglobin that carries oxygen in blood. It also comes in the form of a mineral, magnetite, which can be found in birds' beaks, bees' brains, and fishes' noses. In fact, it's found in the heads of nearly all animals that display prodigious navigational abilities. Magnetite is exquisitely sensitive, aligning with the Earth's magnetic field even though it exerts very little force – around 200 times less than a fridge magnet. Intriguingly, we, too, have magnetite concentrations in the frontal area of our brains, so does this mean that we're capable of magnetoception?

Forty years ago, Robin Baker from the University of Manchester reported some experiments that suggested we could indeed feel the pull of the Earth. He packed blindfolded students onto buses, driving them to the middle of nowhere before asking them to point in the direction of home. As with Burda's experiments with animal alignment, the data were a little noisy, yet 90 per cent of the volunteers indicated a direction that was eerily close to being correct. When all of the guesses were combined, their average was within five degrees, or less than 2 per cent, of the correct bearing. However, when the blindfolds came off and the subjects were asked to guess again, this directional ability disappeared. It seemed that their precision was dependent on them being forced to rely on some ancient magnetic sense; when the dominant sense of vision was added into the mix, it threw everything out. Moreover, those who in addition to the blindfolds had been given magnets to wear on their heads, thus confounding their inner compass, were comparatively hopeless at guessing the way home.

Despite the initial excitement at what seemed to be proof of an extra sense, Baker's findings attracted scepticism. Various research groups tried and failed to replicate his results. Others weighed in to point out that although we may have magnetite in our skull, we lack any kind of sensor that picks up how it responds to the Earth's magnetic field. It was a fluke result; humans don't have magnetoception, case closed. It was a chastening experience for Baker, and one that eventually led him to give up on research, which sadly says a great deal about the sometimes combative and adversarial nature of science.

Since then, tantalising clues have emerged to suggest there might be something in it after all. For instance, there are indications that mammals such as bats do use magnetite as a navigation aid. Meanwhile, fascinating evidence has emerged from imaging studies of the human brain. In 2019, a team of academics led by Caltech's Connie Wang examined the response of human volunteers to changes in the local magnetic field. The volunteers were fitted with electrodes to measure their brain activity and seated like oversized budgerigars inside a cage. This was no ordinary cage, however, but a Faraday cage, which allows experimenters to manipulate the magnetic field within. The results were conclusive: each time the compass was reset, many of the volunteers' brains lit up in response. As well as demonstrating that our brains do register the magnetic field, the experiment revealed two other things. First, none of the volunteers expressed any awareness of what had gone on. And second, there were major differences between brains – some were highly sensitive, while others barely registered the changes at all. Perhaps those among us who seem to possess a superhuman sense of direction are aided by this mysterious sense that operates deep in our subconscious.

*

Aristotle's designation of five senses was based on the intuitive connection that we have with these primary modalities. We're consciously aware of sight, hearing, smell, taste and touch. We can effortlessly but actively entrain these senses, so that we can upgrade from a passive sensory experience to actively engaging them. We can hear, for example, and we can listen, or we can look, and we can see. This isn't always the case with senses beyond the big five. Even while there's evidence that we can detect electromagnetic fields, we're not necessarily aware of them. In this way, magnetoception illustrates the difference between sensation and perception. We seem to share some of the sensory hardware possessed by animals that can recognise and respond to the Earth's polarity, but we don't register it as they do. You could argue that it's down to evolution, which tends to equip organisms parsimoniously, so that only those traits that confer an advantage are selected for. It's obvious that a bird, or a turtle, or any other migrating animal will be well served by an ability to home in on a distant place. For them, it can be the difference between life and death. By contrast, while I'd be pleased to have the capacity to find my way around more effectively, annual migrations have never been part of the human story. Consequently, a built-in biocompass isn't a prerequisite for human success.

While human magnetoception remains a sensory grey area, there are two other modalities that we most certainly do have, but which are often neglected in discussions of our senses. Equilibrioception, our sense of balance, may seem less important than sight, hearing or smell, yet it provides the foundation for an active life. And though we might not be aware of it much of the time, we certainly are when it goes wrong, During the 2016 US presidential election campaign, Hillary Clinton stumbled and fell at an event to commemorate 9/11. Her momentary loss of balance was seized on by her political opponents as a display of vulnerability, a metaphor for her being unbalanced and off-kilter. The episode was pivotal in her eventual loss at the polls a little under two months later.

Maintaining our balance is a team effort from the senses, combining input from our eyes, the sensors in our skin and muscles, and the vestibular system. This latter is an exquisitely formed trio of fluid-filled, looped tubes in the inner ear. As we move, the fluid within these tubes sloshes about and washes across minute sensory hairs that detect the movement and pass the information to the brain. The hairs, and the cells that bear them, are more or less identical to those that are found next door in the cochlea, demonstrating the close anatomical alignment between our senses of hearing and balance. Both came from our ancient fish ancestors, in whom this was a single sense. Having three of these looping, semi-circular tubes, each set at a distinct orientation from the others, allows us to register movement in all three spatial dimensions. Even then, there's a little anatomical finessing, provided by a pair of otolith organs, which are filled with a mass of tiny stones known as otoconia, literally 'ear dust'. Its purpose is to let us know how we're travelling. The movements of the ear dust ultimately tell the brain about our vertical acceleration – in other words, if we're falling – while the other keeps tabs on horizontal acceleration. As you might imagine, both get quite a work-out whenever we go on a roller coaster, or experience turbulence on a flight.

The elaborate compendium of whorls, loops and sacs that comprises the vestibular system render us sensitive to every movement that we make, despite the fact that the entire apparatus is packed into a space no larger than a sugar cube, and the bony tubes with their liquid centres are only a millimetre or so in diameter. It's remarkable to reflect that our balance and coordination stems from this. Walking upright and maintaining our posture throughout our daily activities are both entirely dependent on the vestibular system. The righting reflex, whereby we stabilise ourselves following a stumble, also emerges from some lightning-fast calculations by the brain, based on input from the eyes and the inner ear.

But perhaps the most valuable benefit that the system provides is known as the vestibulo-ocular reflex, or VOR. It's something we don't notice, yet without it life would be far less pleasant. To get an idea just how unpleasant, think back to the last time you watched an amateur filmmaker's home movie. It can be a disconcerting experience, as the jerky movements of the camera yield a result that's entirely unlike our normal perception of the visual environment. The funny thing is that this dismal film is perhaps a truer reflection of the world than the one we see. The seamless vision that we enjoy, even when we're engaged in an activity like jogging or dancing that requires us to move the head quickly and often, is the result of the VOR that keeps our vision on an even keel. It's one of the fastest reflexes that we're capable of, taking less than a hundredth of a second, and involves our eyes automatically moving to compensate for every head movement that we make, in the process providing the smooth visual experience that we take for granted.

In the nightmare scenario that something goes wrong with the vestibular system, even basic actions can become near impossible.

Take for instance the case of John Crawford, a medic from New England, who was put on a long course of the antibiotic streptomycin to combat a suspected case of tuberculosis. While the drug was successful in warding off the disease, it had the unwelcome effect of destroying his vestibular function. After a few weeks of treatment, he described how his hospital room seemed to spin and he suffered overwhelming vertigo. Even to attempt the simple task of reading, he had to wedge his head between the metal bars at the end of his bed because any slight movement, even that caused by the pulsing of his heart, made the words seem to jump sickeningly around the page. He described the effort of walking along a corridor that seemed to reel beneath his feet like the deck of a ship in a storm. Once free of the antibiotics, he began to recover, but even then he was forced to adopt extreme measures to maintain his balance, especially at night when he couldn't calibrate his movements by sight. He relates how, upon leaving an evening drinks party and going into the dark, he would crawl on his hands and knees, his pride compelling him to blame his actions on the effects of a cocktail too many rather than the underlying medical reasons.

For most of us, the input from our eyes and body coordinates with what the vestibular system is telling us. But the moment that the different players of this system start sending contradictory messages, such as when we read in a car, or go on board a ship in boisterous seas, we feel it badly. Why would this be? Humans are obviously highly mobile animals, and our sense of balance has evolved to cope with the normal movements of life. Modern transport, however, provides relatively new experiences for the brain to cope with. The problem arises from the fact that in these situations, we're like Schrödinger's passengers – we're both moving and not moving. In a car, sensors in our muscles tell the brain that we're relatively still, the vestibular system is detecting rocking movements and acceleration, and the eyes are either saying that we're still or that we're travelling quickly, depending on whether we're looking inside or outside the car. In short, it's a mess. The mismatches in these messages cause problems for the brain; since our long evolutionary history has largely been characterised by a lack of cars, it lacks an appropriate frame of reference. Consequently, its suspicions fall back to the one thing that has most often caused these kinds of discrepancies between the senses: poison. The brain knows there's a simple and urgent way to expel toxins, and it makes you throw up.

Sometimes, of course, the reason that we struggle with balance is that we actually have poisoned ourselves, albeit mildly. The cerebellum, a walnut-sized part of your brain that sits near the top of the spine, plays the key role in bringing together the different strands of information we need for balance and coordination. It's also one of the first areas of the brain to be affected by our favourite neurotoxin, alcohol, which is why we stumble and fall over if we've had a drink or three too many. Again, the brain resorts to plan A under these circumstances and makes us vomit.

For all that we relegate our sense of balance to the sensory second division, it's absolutely essential to our quality of life. When people have undergone surgery on the brain, the single biggest predictor of a good recovery is their ability to balance and maintain their posture. That's largely to do with the collaborative nature of balance, which involves input from many different sources and engages many regions of the brain. This is one reason why it takes infants such a long time to integrate all the strands of information to the point where they're able to stand unaided. Balance takes far longer than our other senses to master.

On average, children are a year old before they develop the fine motor control that's required to stand – and even then we call them toddlers because of their accident-prone stumbles. As we age, the vestibular system is prone to wear and tear. Displaced bits of gunk can collect in the semi-circular canals, and the precious ear dust within the otolith organs may diminish. The result is an increase in feelings of dizziness and vertigo in older people, which in turn can lead to falls and debilitating injuries. The good news is that the vestibular system can to some extent compensate for the problems that afflict it. Even so, if you're older than a toddler, and younger than about seventy, you might be guilty of underappreciating equilibrioception; it's the kind of thing you don't miss until it's gone.

*

Even if we allow the sense of balance to take its place alongside sight, hearing, smell, taste and touch, we're far from done. These senses are all what are sometimes referred to as exteroceptive – they tell us about the world outside. Another grouping, the interoceptive senses, tells us what's going on within the body. And sandwiched between the two is a strange sense known as proprioception, or kinaesthesia. It provides us with awareness of the position of our body parts relative to the rest of the body, based on a matrix of sensors that run throughout our joints and muscles. Many people think the idea that you would not know where your own hand is at any point in time is weird. That's not unreasonable; the first place to look is always going to be at the end of the arm. But if this seems trivial, try closing your eyes and touching your nose. The chances are that you achieved this task with something approaching aplomb; for that you have proprioception to thank.

The bodily awareness that proprioception provides is hugely important. We can see it in action when Serena Williams executes a perfect smash, her body and limbs perfectly synchronised to deliver the winning shot. She may not be aware of it, but it's proprioception that allows her to perform in the way that she does. Lesser mortals, such as myself, are still very much in debt to this sense for supposedly simple acts like feeding ourselves or walking down the street.

Sometimes, we can best appreciate our senses by reference to extreme cases, and the story of Ian Waterman illustrates the importance of proprioception. In 1971, Ian was nineteen years old and working as a butcher on the island of Jersey when he was struck with an incredibly rare illness. His doctors were perplexed; what had initially been assumed to be gastric flu appeared to have left him paralysed from the neck down, but there were no obvious causes for his symptoms. His muscles were undamaged; it was the ability to coordinate his movements that had been lost. Later it was discovered that an entire network of nerves within his body had been destroyed, apparently by an extreme autoimmune response.

Baulking at the prospect of spending the rest of his life in a wheelchair, Ian stubbornly focussed his efforts to regain some semblance of his former life. What was required, it seemed, was for him to forge a new link between his mind and his muscles. He found that if he harnessed his other senses, he could begin to compensate for the loss of proprioception. Progress was slow – it took him four months to relearn how to put on his socks, and a year to be able to stand independently. He could no longer take for granted the coherence and organisation that we require for everyday activities. Instead, he must watch his hands as he eats, and his legs and feet as he walks; maintaining a line of sight to his limbs permits him to coordinate their movements.

Although this solution works well enough, it isn't flawless. To this day, without light, Ian becomes helpless. He first discovered this soon after his release from hospital when the power in his mother's kitchen suddenly went off one evening. Ian's newly acquired, sight-based link between brain and body was suddenly cut, and he crumpled to the floor. Just as with all of our senses, if we're lucky enough to never be challenged by their loss, we seldom think of the extent to which we are utterly reliant on them. Ian's road back from illness has been long, yet inspirational. Though he remains reliant upon his other senses to compensate for his loss of proprioception, he mastered the challenge of regaining his quality of life.

*

Proprioception, then, registers the attitude of the body's parts in relation to each other and to the outside world. There's a swathe of other sensors that keep tabs on what's going on deeper within the body. On a recent field trip to the Great Barrier Reef, I dived a few metres below the surface to peer into a submerged cave and came face to face with a huge lobster crouching within. It was hard to tell who was most surprised by the encounter, not least because lobsters have a limited range of facial expressions, but the size of the creature led me to linger a fraction longer than I might have otherwise. Towards the end of this *tête à tête*, my brain's demands to get some air rapidly became overwhelming. Though it was only a short return to the surface, the urgency of the situation was unignorable – taking a deep gulp of air felt like my sole imperative. The reason that my brain was being so insistent likely wasn't because it was short of oxygen – the clamouring is normally due to it sensing a build-up of toxic carbon dioxide and urgently needing to rebalance the situation.

Impulses such as these rise very quickly to the top of the agenda of our consciousness, pushing past the transient charms of colossal lobsters to command complete attention. Of course, there are many other contexts in which our internal state asserts itself. Engaged in a fascinating conversation at a drinks party recently, stretch receptors in my bladder wall began to signal the consequences of my liquid intake. As time passed, what started as mere background noise became increasingly strident, until I had to abandon my companion and take my leave.

Most of the time, however, our behaviour is guided more subtly. For instance, while we might imagine that reaching for a glass of water is an entirely spontaneous decision on our part, the impulse comes at least partly from our subconscious sensitivity to slight changes in the blood's osmotic pressure. In each of these scenarios, the root cause is the ceaseless dialogue that takes place between mind and body. From the beating of our hearts and the chemistry of the blood to the degree of stretch in our stomachs, bowels or bladders, everything is constantly monitored. The brain's perception of what's going on is known as interoception, and it drives our behaviour in any number of ways as the brain seeks to balance the body's physiological accounts.

We often think of the brain as being the commander-in-chief, controlling and coordinating the rest of the body. But though the perception of the body's state does occur in the brain, communication is very much two-way. In fact, since four times as many nerve fibres travel to the brain from the organs than the other way around, it's possible to conclude that the body is in control of the mind. This is never more true than when a biological demand, such as a bathroom break or the need to surface during a dive, becomes imperative. But rather than adopting such a hierarchical view, wherein the mind and body are distinct and separate, it's fairer to say that mind and body are integrated parts of the whole. The intimate connections between them means that our every thought is shaped, to some extent, by what's going on in the rest of the body, and this has profound implications for our emotions, the decisions that we make, and our sense of self.

The realisation of the extent to which the body conditions our mental processes is reflected in the recent surge of interest in the topic of interoception. Some of the most interesting research has examined the question of how attuned people are to their bodies, what's sometimes referred to as interoceptive sensitivity, and a simple test of this centres on our awareness of our heartbeat. Participants are asked either to count their own heartbeats in silence for a period of time, or to listen to a regular beat and determine whether it matches the rhythm of their heart. In both cases, the participants' assessment is based on their consciousness of their heartbeat. You're not allowed to press a finger against a pulse point on the wrist – that's cheating. The results of studies that have used these methods suggest that there's a surprising amount of variation between people in this regard. Something like a third of people are so attuned to their own pulse that they have no difficulty in describing their heart's activity. The remainder of us struggle to a greater or lesser degree with this task.

This finding has interesting and important implications. Specifically, the ability to perceive and understand the messages that come from the body seems to be crucial in how we respond to stress and interact with our emotions. The neuroscientist Antonio Damasio, a pioneer in this field, described how our conscious perceptions are born out of unconscious physiological reactions. When we encounter a stressful situation, our muscles tense and our heart rate quickens. Our emotional and behavioural reactions are based on the brain's detection of these changes; the greater our awareness of our physical state, the more adept we are at interpreting our feelings and responding appropriately.

It's when this dialogue is compromised that problems can arise. For instance, many people suffering with depression seem to have a relatively poor ability to gauge bodily sensations and are consequently out of tune with its processes. Interoceptive awareness appears to be tightly interwoven with our psychology; the same dislocation between the body's signals and the mind are a predictor of obesity, substance abuse and even suicide.

People with anxiety disorders perform a little better at registering what's going on in the body but may misinterpret what's going on. In particular, anxious people may be hyperconscious of small fluctuations in things like heart rate, and the result can be an escalating sense of panic. All these examples might best be understood in the context of interoception as the cornerstone of our sense of self. A fracture in this can have serious consequences, but what can be done about it?

We've heard much about the power of exercise to provide not only physical benefits but mental ones, too. An intriguing idea suggests that such benefits emerge through the ability of exercise to tune up our interoceptive abilities, thereby strengthening the connection between brain and body. Pretty much any exercise will do the job. Strength training, for example, seems to be particularly effective at combating anxiety. Toning the body seems to improve interoception, making us more attuned to bodily signals, and ultimately improving emotional resilience and well-being.

You might not need to hit the gym to realise some of these benefits, though. Techniques such as meditation and mindfulness can play a significant role in attuning mind and body. A recent study by Lisa Quadt and colleagues at the University of Sussex examined the effectiveness of training people to improve their interoceptive ability, specifically their heartbeat detection. As a control, a second group were provided with training that didn't focus on bodily awareness. The results were compelling; those given the interoceptive training not only improved their capabilities in this respect, but also showed a dramatic reduction in anxiety symptoms compared to the control group. The most exciting thing about such studies is not that they show the effectiveness of physical exercise and therapy in reducing things like anxiety, but that we're beginning to understand the mechanisms by which these changes occur. This type of research is still in its early days, but the prospect of developing new methods to target the scourge of mental illness is incredibly encouraging.

Although interoception is a hot research topic at the moment, it has a long way to go before it catches up with what we think of as the primary senses. However, there's an increasing appreciation of the need to mesh our understanding of our outward-facing senses with those that are concerned with the body's inner state. Our overall perception of the world is entirely reliant upon the integration of these. Imagine walking into a bakery first thing in the morning and smelling freshly baked pastries and ground coffee. It's a wonderful sensation, but part of the reason is that you're probably hungry. The wafting scents are so tempting largely because sensors within the body register that your blood sugar is low and your stomach is empty. Visiting the same bakery after eating a big breakfast would be far less of a thrill for the nostrils. The context in which we experience our primary, external senses is set by the perception of our bodily state, which is determined by interoception. Put more scientifically, interoception determines the salience of stimuli from the outside world. It determines which stimuli we should focus on, and consequently shapes our response to these.

This combination of interoception and exteroception is what provides each of us with the perception of ourselves as a single, unified entity; it's at the heart of what we mean when we talk about our sense of self. The rubber hand illusion provides a useful example of this. A participant is seated at a table, with their left arm stretched out in front of them and slightly off to the side, and their hand resting, palm down, on the table. An experimenter then introduces a screen so the participant can no longer see their arm and places a realistic, fake arm, outstretched and flat on the table, just like the real arm. The participant can now see a replica, though perhaps not entirely convincing, arm before them, and at this point they're doubtless thinking that this is one of the most ludicrous things they've ever got tangled up in. As they look at the trick arm, the experimenter simultaneously caresses both the rubber simulacrum and the hidden, real hand gently with small paintbrushes. After a few minutes, the participant is asked to close their eyes and move their right hand below the table until they think it's directly beneath their left hand.

Simple as the experiment is, the results are extraordinary. Around 80 per cent of people report that even though they're 'in' on the trick, they gain the disconcerting impression that the rubber arm is their own. When instructed to move their right hand to meet their left, many position it below the fake hand rather than their real one. What's happening is a sensory confusion between the inputs of vision, touch and proprioception, leading the brain to take ownership of the false arm. Recently, this same experiment was repeated, but this time accounting for the differences between people in terms of their interoceptive awareness. What this revealed was that people who are highly attuned to their bodies tend to be far less susceptible to the illusion. The realisation of the importance of interoception in shaping our perceptions of ourselves and building robust mental health represents one of the most important recent breakthroughs in sensory science.

Our sensory appreciation of the world extends far beyond that yielded by the big five. We might be guilty of underappreciating relatively neglected senses, such as proprioception or balance, but only until one or other of them becomes compromised. At the same time, there's no doubt that interoception colours and contorts our perceptions. The vibrancy of sensory stimuli when we're newly in love, or when we first conquer one of life's challenges, is countered by the muted, or even aggravating, qualities of those same stimuli when we're feeling blue. At the same time, looking away from the lens of subjective experience, and taking inspiration from the sensory experiences occupied by other animals, we can gain a deeper understanding of the world around us.

There are many answers to the question of how many senses we have. It's more than five, perhaps more than fifty. I'd argue that we learn little from the dry arithmetic process of accounting the senses. The important thing is to understand that the end result, perception – our overall sensory experience – is an alloy, an extraordinary conjoining and melding of the separate senses.

# The Weave of Perception

*The brain is the citadel of perception.*

— Pliny the Elder

At the start of this book, I described making my way to engage the minds of a new cohort of students in the wonders of sensory biology on a beautiful spring morning. My routine when I arrive is fairly consistent. As I stride in, giving a smile to the class and endeavouring not to trip up or do anything else similarly Instagrammable, my thoughts are on what I'll say and how I'll say it. In particular, how I'll take the students with me to the summit of sensory biology, the singular, subjective phenomenon known as perception.

In practice, I'm likely to say, as I have done so far in this book, that I'll start by considering each of our senses independently — it's logical to separate the senses out, examining them in sequence to think about how each works. After I've done that, though, I know that I'll have to confront the elephant in the room, perception, and a messy elephant it is, too. While the sensations that we obtain from the outside world and from within our bodies are physical processes, the perceptions that arise from them are psychological. Sensations arriving at the mind are organised, filtered and interpreted, and they're subject to bias and partiality. Our conscious experience of perception thus defies the kind of neat approach that I can use to describe sensations. Indeed, individual perception is so unique, so complex and so nebulous that it sometimes seems to lie outside the purview of science, existing instead at the nexus of art and philosophy. Rather than running away from the messy elephant, however, I hope to confront its disorderly ways and to make it, at least, presentable ...

Collectively, the senses equip us to perceive a swathe of different environmental stimuli. Being armed with wide-ranging sensitivity is incredibly valuable in evolutionary terms, affording us the ability to detect a diverse range of signals that are often transmitted, initially at least, through one medium. For instance, we might smell fire before we see it, or hear the approach of footsteps at night. Having plenty of available channels, each scanning a distinct sensory band, provides an essential breath to our awareness of our surroundings. As well as spreading the risk, being equipped with all of these modalities is what provides our rich, multisensory experience of life. Most important of all, the crucial benefit of having many senses is that it allows us to gain a coherent perspective, the better to understand the world.

Though each of our primary senses operates within its own distinct domain, each is responsive to the others and our perception is a conglomeration of all of them. Paradoxically, we can see how interdependent the senses are by looking at what happens when one is temporarily withdrawn from the mix. For instance, putting a blindfold on a person cuts out vision, which increases reliance on the other senses and also seems to boost their acuteness as the brain reattunes to them. Blocking hearing or touch can lead to feelings of isolation and estrangement from a person's surroundings. Smell also plays an important role in our emotional connection to the world. People who've lost this sense sometimes describe feeling cut off from those around them.

Nevertheless, we sometimes overemphasise particular senses in certain contexts. When we bump into a friend and start a conversation, the dominant sensation is an interplay of sound, of talking and listening. But communication can never be the domain of a single modality; the strongest messages are those that make the broadest use of all of our sensory channels. Nothing made me realise this simple truth more than during one of the saddest times of my life, when I was forced to consider how I could reach someone I loved.

I owe so much to my mum. She was warm, empathetic and incredibly funny. Life was often hard, though, and for much of my childhood, I remember her struggling to make ends meet. As I grew older and branched out on my own, my most fervent hope was that I could find a way to repay some of my debt to her, earning enough to buy some of the luxuries that circumstance had always denied her. My plans never came to fruition; in her mid-fifties she began a decline that was eventually diagnosed as early-onset Alzheimer's disease.

At first, her mental slips seemed innocuous. She'd chide herself, calling herself a 'daft bat' whenever a word escaped her mind's grasp. But beneath the surface the illness was gathering pace. The vividness of her personality began to fade like the petals of woodland bluebells, her favourite flowers, as spring gives way to summer and the forest canopy closes above them. Her descent was marked by awful milestones. First her memory lapses became more frequent and more obvious. Her grasp of reality began to fracture. Then her speech became increasingly confused and she found it harder to comprehend or be comprehended. The sense of humour that we'd shared was lost. The ties that bound us together began to fray, strand by precious strand, until they broke and she seemed to drift away, beyond my reach. Even so, I never fully realised the extent of her separation until the day came when she could no longer remember who I was.

I knew that the mum I'd known was lost to me now and that no amount of wishing could bring her back. Although my visits still seemed to bring her some kind of happiness, the situation seemed to me to be hopeless. I'd arrive determinedly upbeat, but each time my opening sallies failed to spark the response I hoped for. Sitting with her in mute frustration, I tried to think of ways that I could help her. Inspiration came in the form of an article by the late A. A. Gill, who described his own father's excruciating descent into the abyss of dementia. Instead of the desperate futility and painful sadness of one-sided conversations, Gill would take a tub of his dad's favourite ice cream and two spoons and they'd sit companionably and eat it together.

Rather than placing such a reliance on a single sense, I tried to use other sensory stimulations to bring some spark of joy to her life. I still spoke to her, of course, but I expanded my horizons. Copying Gill, I brought her favourite ice cream – she'd always had a sweet tooth. I stroked her hair, brought her favourite old perfume to smell and played her the music she'd known as a teenager. I showed her old photographs, gave her a soft, woollen cardigan and brought flowers for her to enjoy in all their multi-sensory glory.

Not everything worked, but occasionally, fleetingly, her face would light up in a huge smile at some recovered feeling. Following this revelation, I could do something positive in the run-up to seeing Mum, instead of losing myself in the dreadful contemplation of what she, and we, had lost. It was the first time that I'd ever thought about my approach to the senses so broadly, and it changed the way I thought about how we relate to the world and to each other.

One of my changed perspectives related to the way that I, as a trainee biologist, had been taught to think of the senses as being distinct and discrete. Part of this viewpoint derived from the fact that each focusses on a different stimulus, and each one takes its own path to the brain. The ultimate fate of these signals, however, is to be integrated in the experience of perception. So it's in the brain, rather than in the eye of the beholder, or in the ears of an audience, or at the fingertips, where perception happens.

*

I was seventeen when I saw my first brain. Rumours of pickled paraphernalia in the biology backroom had been rife throughout my school career. It was a place of both fascination and fear and we were strictly excluded from it, which only intensified its allure. This biological Aladdin's cave was the preserve of a retired teacher, Mr Beecroft, who frightened and fascinated us pupils in equal measure on account of both his harsh manner and his cultivation of the most unmanageably wild nasal hair I've ever seen. One Friday in my final year of school, my class was unexpectedly beckoned into the backroom. No doubt most of my biology classmates felt like I did, as if we were a troupe of Charlie Buckets being allowed into Willy Wonka's chocolate factory.

Within the vault, ranks of formalin jars were organised with military efficiency along dark, wooden shelves. Contained in these liquid-filled sarcophagi were a parade of suspended specimens, all bleached to a uniform, deathly beige. Retrieving one of the larger jars, Mr Beecroft placed it on the bench. 'The human brain,' he said, gesturing to the whorled mass inside. With what I suspect was a practised flourish, he retrieved the brain and set it on the table before us, where it settled with a faint but disconcerting squelch. 'This was once a person. Everything that made them who they were was in here.'

It was an extraordinary thing to see a brain and to reflect on the idea that this saggy lump of tissue, looking like an overfed walnut, is host to all of our experiences. It seemed implausible. You can take the back off a computer and see the tangles of circuitry and marvel at just how intricate and labyrinthine it all seems, yet a human brain, which is infinitely more complex, appears at first glance to be so simple, almost like a moulded, inert jelly. This strange mismatch between the brain's ability and its appearance seemed to have occurred to many of us, and Mr Beecroft read his audience well. 'Doesn't look like much, does it? But it's the most incredible thing in the universe.' He gestured to a spot on the brain. 'This part here controls speech, this one further back deals with language comprehension.' He pointed out area after area, making the point that the brain is compartmentalised, with each defined region specialising in some distinct function.

This idea of the brain as a modular entity is familiar to any of us who've studied biology, but it's not entirely accurate. As an analogy, imagine you're watching a football match. If you're an aficionado, you can identify the role played by the players, each a virtuoso in their own position. Nevertheless, a striker may be called upon to defend, and in desperate situations, a goalkeeper might go forward to contest a corner in search of an equalising goal. More broadly, the individuals are interdependent, otherwise the team would be pretty hopeless. Thinking of a football team as a cluster of autonomous units is, at best, an oversimplification, and so it is with the brain. So although we can identify neural areas that specialise, one of the reasons that the brain works so well is that it's flexible, allowing for interaction and collaboration across its different regions.

This is never more the case than with the sensory departments of the brain. Each of our primary modalities is associated with a specialised, corresponding area of the brain, known as a primary cortex. Nervous signals from the eye, for instance, travel to the primary visual cortex, while those from the ears are received by the primary auditory cortex and so on. Each of these cortices is separated from the others; partly as a result of this, it was thought that each operated in a self-contained way, processing only the information that they collected from their own sense. In this view, integration of the senses occurred at a slightly later stage, at a higher level of the brain.

We now know that the merging of the senses begins to happen much sooner than this, and that rather than being independent, there's plenty of crosstalk between each of the sensory cortices. The upshot of this interplay is that each of our senses profoundly affects the perceptions of the others. These interactions are common to all of us, but in one group of people they are taken to extraordinary lengths.

*

Some years ago, I was introduced to a fellow student on my grad course. She pondered my name for a moment before saying, 'Ashley ... Oh, that tastes of cabbage.' Putting aside what felt a little like a swipe, I quickly learned that Samara was one of those rare people* who live with the extraordinary condition known as synaesthesia. Her everyday experiences were coloured by unusual sensory interactions. When one of her senses was stimulated, it triggered an additional response, a kind of perceptual ricochet, in an entirely different sense. Samara not only tasted names, she also gained an impression of colours whenever she listened to music. As with nearly all synaesthetes, this incredibly rich, multisensory experience was entirely normal to her. So much so, in fact, that she had been surprised to learn some years earlier that not everyone sensed their surroundings as she did.

For most synaesthetes, the sensory layering and melding is a positive part of their lives, one that augments and textures their sensations. Musicians and artists can gain an additional perspective to that which most of us enjoy, and it's one that assists in creating their art. The jazz composer and band leader Duke Ellington experienced different colours when he heard the notes being played by other members of his ensemble, which allowed him to blend musical hues in his mind's eye according to a kind of sonic palette. In fact, there's quite a list of musicians who experience tone-colour synaesthesia, from Franz Liszt to Billy Joel and from Stevie Wonder to Billie Eilish. It's an ability that's more prevalent among creative people, though it hasn't always been viewed

---

* Estimates variously put the incidence of synaesthesia as low as one in every 2,000 people, or as high as one in every twenty-five.

as a blessing. When the young Vincent van Gogh described his impression of notes having distinct colours to his piano teacher, the instructor thought he was mad and refused to continue the lessons.

Van Gogh described certain musical notes as Prussian blue, dark green, or bright cadmium, but while most of us lack the clarity of this imagery, we do tend to pair unrelated sights and sounds. For example, children learn to associate loud sounds with larger objects before the age of two. Less obviously, we consistently link high-pitched sounds to lighter and brighter colours compared to bass notes. These kinds of associations are so deep-rooted that even our closest animal relatives, chimpanzees, make the same connection. Exactly why we do this isn't clear, though our best guess is that we associate characteristics of the two different stimuli. Loud noises and large objects might match up because they share a similar intensity in our perception. High frequency sounds and brighter colours may go together for similar reasons. If that's true, though, it's harder to explain the link that we make between high-pitched sounds and smaller, spikier objects. Participants in a study run by the pioneering scholar of human perception, Lawrence Marks of Yale University, spontaneously matched high-pitched sounds to an upturned 'V' while associating deeper sounds instead to an upturned 'U'.

Marks' experiment echoes the findings of the psychologist Wolfgang Köhler, who demonstrated another kind of association, one that matches sounds with patterns. Köhler designed two abstract shapes. One was a smooth, blobby outline while the other had a jagged appearance, like an irregular star. Alongside these, he asked participants to match a name to each of the shapes. In his original study, on the Spanish island of Tenerife, the suggested names were Takete and Baluba, while in a later version, for English speakers, the options were Kiki and Bouba. Over the decades since the experiment was first run, participants have overwhelmingly opted to give the spiky, star shape the correspondingly sharp-sounding names Takete or Kiki and called the smooth globule Baluba or Bouba. This naming pattern is chosen by something like 98 per cent of people; it's clear that the words we elect to use to identify visual images are far from arbitrary. It seems to suit our brain to marry the abstract properties of the shapes and the words, putting together a kind of rounded word with the congruent shape and the jagged shape with an appropriately spiky name. In both this example, mapping shape morphology to the sounds of words, and the one connecting musical pitch to brightness, the brain is binding two different yet somehow corresponding sensory attributes. Perhaps it gives those of us who aren't synaesthetes something of the experience of how what we perceive by one sense can percolate through to influence others.

There are other circumstances under which non-synaesthetes can gain an insight into what it's like when the senses bleed into one another. One of these comes courtesy of an invention by a Swiss chemist, Albert Hofmann. In 1936, Hoffman was working on a cereal fungus, known as ergot. For at least as long as people have been growing and eating crops like rye and wheat, contaminated grains had induced seizures and vivid hallucinations, or even gangrene in people who'd eaten goods baked from flour that had been milled using ergot-infested crops. Distinct from such horrors, however, it was thought that ergot showed promise as a source of new pharmaceuticals. Hofmann's aim was to produce a drug that could be used as a respiratory stimulant, but though he painstakingly isolated the active ingredients, his efforts came to little.

Having fallen out of favour, the project languished in the back of Hoffman's mind until a few years later, on a hunch, he decided to reproduce his experiments. This time, he accidentally absorbed a trace amount of the purified material through his skin and went on what was most likely the world's first acid trip. Emboldened by the experience, he deliberately injected himself with a little more of a compound that in his initial work he'd labelled as LSD, and later described the experience beautifully in his journal. 'Kaleido-scopic, fantastic images surged in on me, alternating, variegated, opening and then closing themselves in circles and spirals, explod-ing in coloured fountains, rearranging and hybridising themselves in constant flux. It was particularly remarkable how every acous-tic perception, such as the sound of a door handle or a passing automobile, became transformed into optical perceptions. Every sound generated a vividly changing image, with its own consistent form and colour.' Hofmann was hallucinating, and his discov-ery, which he referred to affectionately as his 'problem child', had somehow transformed him, albeit transiently, into a kind of nar-cotic-baked synaesthete.

Synaesthesia is not, of course, an hallucination, but an authen-tic sensory perception. We know this courtesy of brain imaging studies that conclusively support the testimonies of synaesthetes. Nonetheless, the acid-augmented sensations realised by Hofmann and millions of others since provide some insight into the condi-tion. For decades, synaesthetes were stigmatised as being in some way flawed as a result of disordered connections in their brains. This, we've now come to understand, is a long way from the truth; there's nothing wrong with the mental faculties of synaesthetes. The more modern, neurological take on the difference between synaesthetes and the rest of the population is more to do with the structure and excitability of the former's brains; they have a greater abundance of connections between their sensory cortices, meaning that activation of one sense stimulates a strong and vivid response in others. One of the most compelling suggestions is that during the earliest stages of our lives, indeed before we're born, different sensory regions of the brain are extensively connected. As we continue to develop, these links are progressively pruned until all that remains are isthmuses of neural tissue spanning adjacent brain areas. The effect of this is to maintain connections between the senses while limiting the amount of crosstalk. In synaesthetes, however, this paring-back is proposed to occur to a far reduced extent, meaning that sensory cortices are hyper-connected. The flow between distinct modalities is therefore so much the greater, in turn making many experiences far richer.

To a lesser extent, this sensory cross-over exists in all of us. A study by Jahan Jadauji and colleagues at McGill University and the University of Pennsylvania found that when activity in the visual cortex of non-synaesthetes is triggered experimentally, in this case using a magnetic field, there's a corresponding response in the olfactory cortex. The net result of this procedure is that the participants' sense of smell can be piqued, making it easier for them to distinguish between odours. The same sensory collaboration also works in the other direction, so that the visual cortex as well as the olfactory centres are activated when we're sniffing something. When you smell a lemon, chances are that an image of the fruit might pop into your mind's eye. So rather than compartmentalising the senses in the way that we once thought, the brain seems to actively recruit multiple modalities as a means to gain a broader perspective on a stimulus.

Synaesthesia can potentially emerge through the connection of any two sensory pathways. The most common form is what's known as grapheme-colour synaesthesia, where letters and numbers are experienced as distinct colours. For instance, the letter E might be pink, and the number 3 bright yellow. The concept of seeing monochrome numbers and letters in glorious technicolour is sometimes dismissed by sceptics as merely the product of associations formed in early childhood when many of us learn these figures using brightly coloured symbols. There's strong evidence, however, to indicate that this isn't so. A sheet of paper covered in numerals might look meaningless to most of us, but a synaesthete can effortlessly pick out hidden patterns. Let's say that within the cloud of jumbled figures, number 5s have been placed so that they form the corners of a square, or 8s are positioned at each point of a triangle. Non-synaesthetes won't typically see this among the mass of digits, but a true synaesthete will see it right away. And if as they do this, they're wearing apparatus that enables us to see what's going on inside their brain, we'd see that a region of the visual cortex that responds strongly to colour becomes highly activated upon seeing monochrome letters or numbers. The activation of this region on viewing these figures occurs just as quickly as it does when it gains colour information direct from the retina.

Since synaesthesia has the potential to mingle any of our senses, there may be as many as 200 different forms that it may take. There are people, for instance, who perceive tastes while touching objects; shaking hands with someone might induce a bitter flavour in their mouth that persists for hours. In other synaesthetes, a taste might generate a tactile sensation. One such person described how foods containing cinnamon gave him the feeling of running his hands through sand. Two things seem to be consistent in true synaesthetes. First, their responses are consistent; the person whose fingertips seem to be transported to a beach upon eating cinnamon will experience the same thing every time he eats cinnamon. And second, the response of the induced sense doesn't usually match up in any obvious way to the initial perception – there's no apparent reason that shaking hands with someone should deliver a bitter taste. What the brain of a synaesthete is doing in these instances is connecting apparently unrelated sensory perceptions. This trait is a fundamental part of the way in which the human mind allows some people to link ideas and concepts, to see the world differently, to be creative. More broadly, working out how and why the brain connects different stimuli to arrive at its perception of the world lies at the heart of our understanding of what makes us tick. Seen in this light, synaesthesia may offer an insight into the evolution of the human mind and its most extraordinary manifestation: consciousness.

*

By working together, the senses compensate for situations when the information from any single sense is patchy or weak. Among a hubbub of chatter in a crowded room, it can be a constant challenge to follow a conversation if you're relying entirely on sound. One option is to turn your head slightly to point an ear in the speaker's direction, but the greatest value comes from watching the speaker's face. Studies of people in such situations show that as the background noise levels increase, listeners tend to spend more time training their gaze on the mouth of the speaker and less time making eye contact. In this way, vision is increasingly recruited to make up for the difficulties faced by the ears. Neither sense, on its own, is capable of providing the listener with a full understanding; together they combine in what's known as a superadditive way to elevate our comprehension.

The ability of the brain to integrate sensory inputs depends on how well those different signals match up. In a conversation like the one described above, the alignment of voice and facial movements makes it simpler for the brain to assimilate the two into a clear perception. The more asynchronous or spatially separated, the less likely we are to gain a coherent view. When a soundtrack and images of a video don't link up properly it's not difficult to work out what's going on, but the whole experience feels jarring and consequently harder to enjoy. The brain does, however, allow a little leeway. Something similar happens at the cinema, where the speakers are usually offset to the side of the screen. Instead of localising the audio and the video separately, the brain superimposes the former onto the latter to give the impression that the sound actually comes from an actor. This is again an example of the brain creating its own subjective version of reality, rather than reflecting the objective truth.

We can get an idea of how the brain matches up different senses from an illusion where a single flash of light is presented alongside two rapid pulses of sound. What usually happens is that the sounds gull us into believing that there have been two flashes of light rather than one. It's surprisingly convincing and more evidence of the brain's jiggery-pokery in this regard, but perhaps the most interesting aspect of it is that just how compelling the illusion is depends on how close the pulses of sound are in time. It seems that if they're a tenth of a second apart, or less, the brain provides the perception of two flashes of light, but if the interval is longer, this doesn't happen. This suggests that there's a short window of time that the brain uses to determine if different sensory streams are linked and, fascinatingly, a further suggestion that the length of this window is at least partly determined by the rhythms of activity, known as brain waves, that sweep across our grey matter about ten times a second.

When two or more of our senses combine, that message is stronger as a result. In tests of reaction times, people typically respond far more quickly to a light flash and an audible beep when they are presented simultaneously. Whenever I'm lecturing, I try to bear in mind some advice I received early in my career that I should speak with my body as well as my mouth. By gesturing with my arms in a way that's in tune with what I have to say, I can get my point across more effectively. It's easy to tell when I'm at the front of a class whether the students are engrossed or have slipped into a cataleptic state, and gestures play an important role in making sure it's the former. As well as augmenting the content of my lectures, gesturing tends to draw attention. This mental entrainment further engages the senses, so a positive feedback loop develops; multisensory cues capture our attention and in turn increase our concentration on a stimulus.

If I was to look closely at the students' faces in my lecture, there's one absolute giveaway to their level of engagement – the size of their pupils. These increase in size when we first detect a particularly salient stimulus, but if we're focussed on an ongoing, challenging task they tend to be intermediate in size. I'd like my students' pupils to reflect this as they gobble up morsels of interest from the smorgasbord that I'm laying before them. The problem is that to get the definitive answer to the question of what their pupils are doing and thus how engaged they are, I'd have to loom over them in a way that would make their pupils dilate to the size of buttons. In any case, intense focus often waivers over time. This isn't so important in a lecture, but in other contexts zoning out can be disastrous. Lapses of concentration seem to occur most often when the pupils are either dilated or constricted; in between these is a sweet spot where the brain is in a medium state of arousal, during which it's least prone to suffer from attention deficits.

Attention can also cause us to promote our focus on one sense over all of the others. Disturbed by an unexpected noise in the dark of night, we strain to listen. As we do so, the brain automatically responds to our demand, scaling down the activity in some sensory processing regions while boosting its gain for hearing. Imaging studies carried out on volunteers subject to changing stimuli show that as they attend to new information and switch their attention, there are changes in the patterns and locations of electrical animation in the brain. Being able to entrain our sensory attention is vital for establishing the levels of concentration needed for tasks such as driving, but switching attention exacts a cost. Changing the brain's attentional gears involves a slight lag in our sensory processing. The buzz of the mobile phone momentarily draws our interest and our attention to the road evaporates. In that time, the likelihood of an accident increases massively. Even speaking on the phone while driving elevates the risk by a factor of around four.

The brain's flexible approach to allocating its resources according to the sensory demands made of it can be seen most clearly among people who lack one or other sense. For instance, deaf people gain improved vision, while the blind benefit from superior hearing. In these cases, the brain seems to recruit additional processing space from whichever of the primary cortices is being underused. Studies of this phenomenon, known as neural plasticity, show that when blind people read Braille or listen to sounds, areas of their visual cortex get in on the act to support the effort. The brain hence makes intelligent decisions about allocating processing efforts, based on what's available. This flexibility of the brain occurs even among sighted people who've been deprived of vision for a short period. When someone has been blindfolded, for instance, it doesn't take long for other senses to hitchhike those areas of the brain that are idling, making use of computational vacancies to boost the perception of those modalities.

*

The work that the brain does in creating a robust version of reality is reliant on its ability to rapidly resolve conflicts, such as can happen when the brain receives mixed messages from the senses. For instance, most people have had the experience of being stuck in a traffic jam with other cars alongside or sat on a train that's stopped at a station. If a neighbouring car starts to edge forward, or another train pulls alongside, it can feel as though it's us that's moving. If we're driving, it can feel unnervingly as though we're about to roll into the car behind. The world is full of mixed sensory messages that have the potential to trick the brain, yet most of the time it resolves these so effectively that we're seldom aware of the knots and tangles in the information streams that pour into our conscious awareness. To do this, it cross-references different streams in order to arrive at a solid perception of what's going on. In the self-motion illusion described above, the visual information is initially compelling, but out of sync with other sensory stimuli. The vestibular system with its capacity for detecting movement rides to the rescue, countermanding the optical illusion and reassuring the brain that we're not moving at all.

Most of the time, to work efficiently in ironing out these ambiguities, the brain relies on simple rules. One such rule is to weight the input from some senses more heavily than others, so that we end up with a sensory hierarchy, an idea that Aristotle – who else? – proposed over 2,000 years ago. In his scheme, vision was the dominant sense, followed by hearing, smell, touch and taste. Though the primacy of vision and hearing over the other senses seems unassailable, like many such ideas, it's open to question.

Is Aristotle's sensory hierarchy really the natural order of things? Recently Asifa Majid of the University of York and her colleagues set out to test this contention, delving into diverse cultures around the world to find out if other people sense things differently. One ingenious way of testing this is to examine language; after all, the breadth of the descriptive vocabulary should reflect the relative importance of each sense to its speakers. Sure enough, in English and most other Western tongues, the number of words associated with each modality matches closely the order proposed by Aristotle; a huge number of words are devoted to visual perception while only a smattering are concerned with smell. But casting the net a little wider changes the picture dramatically. Among Farsi, Cantonese, Lao and a plethora of Central and South American indigenous-language speakers, the most expansive lexicon is reserved for taste. In West Africa, the bias is in favour of touch, while the Indigenous Australian language represented in the study placed smell as the dominant sense. What this tells us is that, contrary to what we've long assumed from a Western perspective, there is no universal human hierarchy of the senses. Instead, our overall multisensory experience is extensively shaped by our culture.

You can see evidence of the sensory hierarchy, at least among Westerners, in what's known as the McGurk effect. In a 1976 study entertainingly entitled 'Hearing Lips and Seeing Voices', Harry McGurk and John MacDonald stumbled across this strange phenomenon while engaged in the study of infants responding to their mothers' voices. They discovered that when a video of someone speaking the syllable 'ga' was overdubbed with the soundtrack of someone saying 'ba', they each heard 'da'. Initially they thought it was a quirky mistake, but they soon realised that what they'd uncovered was evidence of how the brain sorts rival information streams. When we're in conversation with someone, we not only hear what they're saying, we also process speech information by reading their face. When someone says 'ba', they typically adopt a different facial expression from that which they'd use when they say 'ga'. In this case, the brain perceives a conflict: what it registers from the lips of the speaker doesn't match the sound input. It resolves this by biasing toward vision, as per Aristotle's hierarchy, so that what's seen in the speaker's face overrules what it hears.

Since the original study, many other variations have been tried, including on speakers of a range of different languages. The original pattern holds for European languages, but Japanese and Chinese listeners are far less susceptible to it, again suggesting that the way our brains organise and perceive the world is sufficiently malleable to be strongly influenced by culture.

For most of us, the closest we get to the realisation that the brain is artificially shaping our experiences is when we experience sensory riddles like the McGurk effect, or the rubber hand illusion that I described earlier. There are a quite a few similar multisensory experiences that also challenge our conceit that what we experience is objectively real. For instance, one particular illusion uses sound to twist our perception of touch. Participants don a pair of headphones and then rub their hands together in front of a microphone, which relays the sound to their waiting ears. By tweaking the audio, experimenters can dramatically change the touch sensation that the participants receive from their hands. A roughened sound makes them feel as though their hands are as dry as tree bark, while a smoothed sound fools them into thinking that their hands are supple and soft. Although the participants are in on the trick, their tactile experience seems real to them.

The perceptual flux caused by integrating different senses shapes our experience in each. A swathe of fascinating studies have examined how a change to one sense can affect our perception in another. In each case, participants are asked to report on their experience based on one sense while the background against which they perform a task is manipulated. For instance, ambient smells affect our tactile experiences. Describing the feel of material, people assess it to be softer when they smell a waft of lemon than when they're exposed to a less pleasing scent. Sounds also influence our perception of touch; an electrical toothbrush with an artificially augmented buzz feels harsher than one with a regular hum, even though the toothbrushes are identical. Asked to taste two gels that were identical apart from their texture, people consistently characterised the softer of the two as being far more flavourful. Visual cues, too, influence our perception of flavour. In particular, brightly coloured foods seem to have a far more intense taste than their more drab equivalents.

Emotions and moods further shape our senses. When suffering from anxiety, we're less likely to notice salty or bitter tastes, while negative emotions make us less perceptive of subtle odours. Even posture interacts with the other senses. In part this is because it too is informed by our emotional state, and in turn our entire sensory viewpoint, but it can also change our perspective more directly. In one fascinating study, conducted in Paris, people leaning to one side reported how when they did so, the Eiffel Tower looked smaller.

Meanwhile, thanks to something known as the motion bounce effect, sound can change what we see. When two discs travel across a computer screen on intersecting paths, without any sound, they seem to cross over and go their own separate ways. Adding a sound to coincide with the point at which they meet, however, changes the whole thing: the discs seem to rebound sharply off each other like balls on a pool table. The two versions of the process are identical, except for the sound, yet we interpret them completely differently; depending on which senses are recruited to resolve the problem, we experience a cast-iron certainty that one version is authentic and the other a falsehood. In this, as in all the previous examples, when presented with a series of conflicting scenarios, the brain effectively calculates the likelihood of each being correct and shapes our perception decisively in favour of one. It does so in a way that we now realise is akin to Bayesian statistics, a system that weighs the probability of events based on prior expectations and assumptions. It's a sophisticated method that minimises, but cannot eliminate, the possibility of errors, simply because of our biases. So, for much of the time, the brain is pretty accurate in its verdicts, but even so, it can only ever provide a version of reality.

Working out what's real and what isn't can be tricky, a problem shared by the jewel beetle. The males of the species are romantic wanderers, taking to the wing in search of passion. In contrast, the females live a less nomadic life. Unlike the males, they can't fly and instead wander the countryside of Western Australia, stopping at flowers to feed on nectar. When one of the winged lotharios happens upon a female, he alights and delivers whatever the beetle equivalent of a chat-up line is. However, an increase in littering changed the males' priorities in a rather dramatic way.

Among the litter were what Australians call stubbies, beer bottles made of stippled brown glass. Who can say what goes on in the mind of a beetle, but it's clear that airborne males spotting a stubby – roughly the same colour as a female beetle – experienced something akin to love at first sight. Although the stubbies are vastly larger than their more traditional choice of mate, the prospect of a gigantic paramour only seemed to inflame them further. With a stubby in view, the male would swoop down and begin a fruitless attempt to mate with the bottle.

So intent were the males on consummating their unrequited love that many would be physically dismembered by ants while they were *in flagrante delicto*. Rather than this acting as any kind of discouragement, another lusty suitor would take his fallen comrade's place, only to suffer the same fate. The male beetles' infatuation with beer bottles shows just how susceptible their perception is to error and how this can transform their reality into something quite bizarre. To the beetles, the bottles are what's known as a supernormal stimulus, an exaggerated version of an existing stimulus that produces a correspondingly exaggerated and intense response. But don't imagine that it's one rule for beetles and another for us; we're prone to these kinds of ramped-up stimuli, too. The junk food industry, for instance, delivers massive hits of salt and fats that make their products irresistible to some. Plastic surgery, and particularly the augmentation of lips, breasts and buttocks, taps into pre-existing biases to amplify sexual attractiveness. Immersive online games and targeted advertising use brighter, more dramatic simulacra of reality to appeal to us. We might be equipped with a much more sophisticated brain than a jewel beetle, yet such illusions and stimuli demonstrate just how mutable our perceptions are and how illusory are our realities. And this doesn't just affect our experience of the present, it's written through our past, and in particular our memories.

*

The neuroscientist Wilder Penfield, whose work led to the development of the cortical homunculus, made a host of discoveries during his explorations of the brain. Perhaps the most dramatic of these revealed itself in the 1930s, when he was perfecting a technique that later became known as the Montreal Procedure. His goal was to try to treat patients with severe epilepsy, and to achieve this the patient would lie awake under local anaesthetic, while Penfield probed their exposed brain using weak electrical stimulation. Unexpectedly, when he touched the temporal lobes of some of his patients, they would describe vivid past experiences, many of them long forgotten. Moreover, when he touched precisely the same spot again, the experience would replay.

The experiences themselves were diverse, but the recall was incredibly detailed. They ranged from the mother who suddenly found herself back in the delivery suite, giving birth to her child, and the young boy who saw himself laughing with his friends in the garden, to the girl recalling an orchestra she'd heard, humming along to the tune that had resurfaced at the forefront of her consciousness. It was as though Penfield had discovered the distinct location of memories within the brain. To the patients, however, it was somehow different to simply seeing a past event in the mind's eye. Many of these patients actually felt as though the scenario was being relived. For just that moment, when Penfield's electrode touched their temporal lobe, their perception was shunted from the immediate present, to an event that had already happened.

And what of people with mental conditions that cause them to experience hallucinations? For instance, we're told that schizophrenics interpret reality abnormally, but their experience is as real to them as ours is to us. When a person's perception of reality disagrees to such a great extent with the general consensus it can be convenient to think of it as an aberration, as an outlier that tells us little about how most of us experience the world. In that case, consider eyewitness testimony, which for centuries has been a foundational part of the legal process.

The faith that we place on eyewitness accounts is exemplified by a study in which two mock trials were conducted, identical in every way except that in one there was an eyewitness and in the other there was not. Despite the evidence being the same in each case, the existence of a witness who claimed to be able to identify the culprit led to a fourfold increase, from 18 per cent to 72 per cent, in the likelihood of jurors arriving at a guilty verdict. The supreme confidence that we invest in such eyewitnesses, however, is misplaced. In the US, almost three quarters of wrongful imprisonments that are ultimately overturned on the basis of new evidence were originally convicted primarily on the basis of inaccurate eyewitness accounts. In 2012 Lydell Grant was convicted of the murder of a man outside a nightclub in Texas, with six independent eyewitnesses having identified him as the offender. It wasn't until 2021, nine years after his trial, that new evidence exonerated Grant and exposed the real murderer. These eyewitnesses, like the overwhelming majority of those in other cases, didn't set out to mislead; they firmly believed in their own recollection. The reality, however, was different.

In a less public way, most of us have known episodes like this. You might be discussing some shared event with a friend, only to find that your chum remembers it in a fundamentally different way. Both of you experienced the same thing, yet your perspectives diverge. Yet even with this clear-cut demonstration of how capricious our perceptions are, most of us tend to trust our memories implicitly. It's said that every time you recall a memory, you're remembering the last time you remembered it. As this happens, details become jumbled, displaced and shaded by context. The divisions between you and your friend start out in the idiosyncrasy of your individual perceptions and continue to deviate as they assume their places in the shrubbery of your respective minds.

*

The coexistence of many different perspectives and perceptions is one of the most fascinating things about the senses, and it's something that thinkers have long contemplated. Take, for instance, a refreshing walk in the countryside.

Invigorating as such things are, such a walk brings with it the possibility of an encounter with some small creature that's after a mouthful of your blood. I don't know if you've ever had cause to wonder what it's like to be a tick, but that's one of the unlikely world views chosen by the German biologist, Jakob von Uexküll.

Perhaps von Uexküll chose the tick because they are simple animals, characterised by what can only be described as an incredibly high boredom threshold. An adult tick sits patiently on a frond of bracken, or the edge of a leaf, arms raised in the manner of a cabaret singer in full flow. Hours, or more likely days, go by and the indefatigable tick remains at its station, legs akimbo, ready to embrace any passers-by. Though they might look like they're doing nothing, they're sniffing the air in the hope of a waft of butyric acid, a tell-tale ingredient of mammal sweat. This primes them for action and when the target gets closer, the tick can use body heat and the vibrations caused by movement to home in. If it's lucky, the parasite can latch on to the victim, find a bare patch of skin and delve in delightedly until its entire head is buried in its bloody buffet.

Although ticks aren't the most inspiring of animals – you don't see children, wide-eyed with excitement, crowding and jostling around the tick enclosure at the zoo – the point that von Uexküll was making is just how different the perceptual world of a tick is to ours. During our our bucolic stroll, we might take in the scenery, stop to smell some flowers, and listen to some melodious birdsong. These are all as nothing to the tick, whose sensory existence amounts to catching a waft of sweat, tracing it to its origin, and then, if it's lucky, feeling its way to a meal. The tick's *umwelt*, a term coined by von Uexküll to describe a unique, subjective sensory world, is limited to its appreciation of a vanishingly small number of relevant signals. To us, the tick's perceptual experience seems hopelessly limited. To the tick, perhaps, it seems rich beyond measure.

Each different species has its own *umwelt*. For instance, dogs live among swirling chemical currents of smell and choruses of sounds that we're largely ignorant of. Though there is overlap between our *umwelt* and that of our pet pooch, the fact remains that we experience our surroundings in very different ways, and this shapes not only what we perceive, but how we relate to the world.

Although Von Uexküll was primarily interested in understanding variations between species with his concept of the *umwelt*, there are also considerable disparities within species. Each individual animal exists in its own sensory universe, parallel with, yet distinct from, its fellows. That's never truer than in our own species. Every moment of our waking lives is enlivened by a cornucopia of sensory stimuli. These are intercepted by our sense organs and sent to our brain where perception is created. Since no two people are exactly alike in respect of either their sense organs or their brain, every perception is relative and subjective, so everyone lives in their own *umwelt*. Consequently, your internal representation of the colour red, the taste of bread, or the sound of Beethoven is unknown, and unknowable, to me; I can only construe my own perceptual experience. Philosophers sometimes describe these sensations, the way things seem to us, as qualia. These are part of our personal, idiosyncratic consciousness. Moreover, I can't adequately convey my qualia of red, bread, or Beethoven. They shrink from objective description, and force me to rely on comparative, second-hand and often tautological depictions, so preventing anyone from gaining complete insight into my sensations of them.

Not that it stops people from attempting to describe their qualia. Some of the most famously entertaining examples come from wine experts, who in the absence of any existing framework of language that might render understanding, plunder the rest of the sensual world for analogies. A wine might be enthusiastically likened to hedgerow fruits, ripening hay and toasty oak, all of which are vague and essentially meaningless. The wine experts are doing their best, but I sometimes think they might just as well stick pins at random in a book of metaphors. But if this is the case, why do we bother? It's because our sensory experiences are what makes life worth living, they're the basis of our conscious selves; as social animals, we're predisposed to share our perspectives and to be interested in those of others.

The sensory variations between us give rise to different viewpoints, outlooks and attitudes. They provide a richness and diversity in society. Imagine if we were all equipped with identical perception. We'd lose our individuality, and the world would be a far drabber place. The differences between us come from the genetic architecture that structures our sense organs and moulds our brains. They also come from our environment and personal experience, helping to determine which stimuli we find salient and pay attention to, and the way that incoming sensory information is molded by our prior knowledge and our outlook. And they come from our culture, shaping our expectations and our biases. The limitless permutations of these mean that we're each sensorially unique. Your perception is not just different from mine, it's different from that of anyone who has ever lived.

But although your perception is different from mine, we have much in common. Visiting a supermarket, we might be struck at first by the huge number of choices that we're confronted with. We might get the same feeling when shopping online for clothes, or books. Broad though the options may seem, though, they're actually remarkably constrained. According to Edited, a US company specialising in retail analytics, over a third of the clothes we buy are black. Add in grey and white, and we've already accounted for over half of all the colours that we wear. Next in the list is navy blue, another slightly underwhelming choice. A vast array of other colours exists and yet, in clothing choices, they exist very much at the margins. Our food choices are scarcely any more inspired. Of something like 80,000 species of edible plants, we cultivate only around 150. This narrows further so that thirty of these crops contribute to 95 per cent of human calorific intake worldwide. Indeed, just four plants – rice, potatoes, maize and wheat – provide over half of what we eat. This same pattern, a strong skew towards conformity, guided by our senses, operates in just about all of our perceptions and ultimately our choices. That's another aspect of humanity as a social species: we tend to align with and emulate each other. We might each be different, but we're not so different.

Perception arises from a network of mutually dependent senses that together act as an interface between ourselves and the rest of the world. While each of our senses are, at least in physiological and anatomical terms, distinct from one another, as they reach the brain, the boundaries between them blur and they merge into a singular perception. Put another way, the senses are nothing without each other. If we were restricted to having just a single sense, most people would pick vision. But if you'd been born with vision alone, you'd struggle to comprehend the world. That's because early in our development, accurate perception requires that vision is benchmarked against our other senses so that we can understand what our eyes are telling us. Babies handle objects as they look at them and, as they do so, tactile cues bring meaning to what they can see. Similarly, they taste things, smell them and listen to them.

In those first few months of a baby's life, haphazard fumblings provide the foundations for a powerful collaboration between their senses that will serve them for the rest of their life. Inside their brains, separate sensations are arranged, aligned and blended. But at this early stage, the senses are busily organising the infant brain. Neural connections are being built and strengthened to provide a framework for understanding the outside world. In time, prior experience and expectations, as well as emotion, will be used to temper the raw input from the senses. The result, an individual's perception, will be unique, idiosyncratic and subjective. It won't mirror the outside world, but will represent the brain's best guess and it'll be an imperfect representation. Yet for all that, our senses, bound together into perception, provide the most incredible, extraordinary experience possible, the conscious sensation of being alive.

# Afterword

Back in the lecture theatre, my allotted hour is almost up. In that time, I've spread an array of weird and wonderful material before the students. I've explained that there's no such thing as colour and described how plants can 'hear' approaching caterpillars. I've told them about goats foretelling volcanic eruptions and related how iPhones reconfigure their brains. From their expressions, I can tell I'm getting through, and no wonder – it's the most compelling topic in biology. Then again, I've been at pains to point out that this isn't solely about biology. Just as perception emerges from the confluence of many different modalities, so a deep understanding of this topic is reliant upon a synthesis of many different fields of research; no one discipline can claim ownership of the senses. Biology can be used to examine many aspects of sensation, such as the way that the students now watching me do so with eyes whose evolutionary journey began billions of years ago as light-sensitive proteins in microbes, and listen to me with ears bequeathed to them by ancient, pressure-sensitive fish. It can also probe the relationships between our human senses and those of other animals to explain how it is that we ended up with our sensory world view.

But the importance of the senses to our everyday lives isn't a question that biology alone can answer. Our attitudes, emotions and health are as much a part of perception as the anatomical structures that collect data on the outside world. At the same time, the way in which sensory stimuli govern our behaviour, making us more likely to form an opinion, buy a certain product, or even recover from illness, are topics on which biology has little to say. So biology must be aligned with psychology, philosophy, economics, engineering and medicine, among others, if we are to fully appreciate and ultimately to realise the enormous potential of our sensory selves.

What does the future hold? There's the tantalising prospect that we'll be able to engineer advanced bioinspired sensors that will allow us to harness the full power of the senses. For instance, the eyes of mantis shrimps have inspired the development of imaging technology for medicine and remote sensing that allow an enormous upgrade on what was previously possible. And before long, it may well be that chemosensors, electronic simulacra of our own noses, will be fitted as standard in smartphones. Although this might conjure the idea of Siri hinting gently that, based on our odour, we might benefit from an overdue shower, the opportunities that this would afford in early diagnosis of a swathe of conditions are incredibly exciting.

But while technology races ahead, there's still much to resolve a little closer to home. We know the basic mechanisms of the senses; we know, for instance, that photons of light stimulate cells in the retina or that molecules in the air are apprehended by specialised receptors in the nose and that these are converted by the process of transduction to signals that the brain can understand, but we don't know why those signals are interpreted as they are. The brain's truly awesome ability to create meaning from these signals remains an enigma.

Perception is the key to the greatest and most fascinating mystery in science. How does a biological system such as a human being parse significance from the tumult of physics that surrounds us? Can we ever figure out the relationship between the objective reality of the world beyond us and our subjective experience of the same? Solving the conundrum of perception would lay open the road to understanding the greatest phenomenon of all, that of consciousness.

# Afterword

Many fascinating questions remain and the coming years promise exciting journeys of discovery. My lecture finishes and the students file out of the theatre, chatting animatedly about what they've learned. I can only hope as I watch them leave that perhaps it'll be one of them who finds the answers.

# Further Reading and Selected References

## What the Eye Sees

Alvergne, A. et al (2009). Father–offspring resemblance predicts paternal investment in humans. *Animal Behaviour*, 78(1), 61–69.

Beall, A. & Tracy, J. (2013). Women are more likely to wear red or pink at peak fertility. *Psychological Science*, 24(9), 1837–1841.

Caves, E. et al. (2018). Visual acuity and the evolution of signals. *Trends in ecology & evolution*, 33(5), 358–372.

Cloutier, J., et al. (2008). Are attractive people rewarding? Sex differences in the neural substrates of facial attractiveness. *Journal of cognitive neuroscience*, 20(6), 941–951.

Ebitz, R., & Moore, T. (2019). Both a gauge and a filter: Cognitive modulations of pupil size. *Frontiers in neurology*, 9, 1190.

Fider, N., & Komarova, N (2019). Differences in color categorization manifested by males and females: a quantitative World Color Survey study. *Palgrave Communications*, 5(1), 1–10.

Fink, B. et al. (2006). Facial symmetry and judgements of attractiveness, health and personality. *Personality and Individual differences*, 41(3), 491–499.

Gangestad, S. et al. (2005). Adaptations to ovulation: Implications for sexual and social behavior. *Current Directions in Psychological Science*, 14(6), 312–316.

Irish, J. E. (2018). Can Pink Really Pacify?

Jones, B. et al. (2015). Facial coloration tracks changes in women's estradiol. *Psychoneuroendocrinology*, 56, 29–34.

Kay, P., & Regier, T. (2007). Color naming universals: The case of Berinmo. *Cognition*, 102(2), 289–298.

Little, A. et al. (2011). Facial attractiveness: evolutionary based research. *Philosophical Transactions of the Royal Society B* 366(1571), 1638–1659.

LoBue, V., & DeLoache, J. (2011). Pretty in pink: The early development of gender-stereotyped colour preferences. *British Journal of Developmental Psychology*, 29(3), 656–667.

Maier, M. et al. (2009). Context specificity of implicit preferences: the case of human preference for red. *Emotion*, 9(5), 734.

Oakley, T. & Speiser, D. (2015). How complexity originates: the evolution of animal eyes. *Annual Review of Ecology, Evolution, and Systematics*, 46, 237–260.

Palmer, S. & Schloss, K (2010). An ecological valence theory of human color preference. *Proceedings of the National Academy of Sciences*, 107(19), 8877–8882.

Pardo, P. et al (2007). An example of sex-linked color vision differences. *Color Research & Application*, 32(6), 433–439.

Provencio, I. et al. (1998). Melanopsin: An opsin in melanophores, brain, and eye. *Proceedings of the National Academy of Sciences*, 95(1), 340–345.

Rodríguez-Carmona, M. et al (2008). Sex-related differences in chromatic sensitivity. *Visual Neuroscience*, 25(3), 433–440.

Ropars, G. et al (2012). A depolarizer as a possible precise sunstone for Viking navigation by polarized skylight. *Proceedings of the Royal Society A*, 468(2139), 671–684.

Scheib, J. et al (1999). Facial attractiveness, symmetry and cues of good genes. *Proceedings of the Royal Society of London B*, 266(1431), 1913–1917.

Shaqiri, A. et al (2018). Sex-related differences in vision are heterogeneous. *Scientific Reports*, 8(1), 1–10.

Shichida Y, Matsuyama T. (2009) Evolution of opsins and phototransduction. Philosophical Transactions of the Royal Society B, 364(1531):2881–95.

## Hear, Hear!

Altmann, J. (2001). Acoustic weapons – a prospective assessment. *Science & Global Security*, 9(3), 165–234.

Appel, H. & Cocroft, R. (2014). Plants respond to leaf vibrations caused by insect herbivore chewing. *Oecologia*, 175(4), 1257–1266.

Conard, N. et al. (2009). New flutes document the earliest musical tradition in southwestern Germany. *Nature*, 460(7256), 737–740.

Deniz, F. et al. (2019). The representation of semantic information across human cerebral cortex during listening versus reading is invariant to stimulus modality. *Journal of Neuroscience*, 39(39), 7722–7736.

Ellenbogen, M. et al. (2014). Intranasal oxytocin attenuates the human acoustic startle response independent of emotional modulation. *Psychophysiology*, *51*(11), 1169–1177.

Ferrari, G. et al. (2016). Ultrasonographic investigation of human fetus responses to maternal communicative and non-communicative stimuli. *Frontiers in psychology*, 7, 354.

Fitch, W. (2017). Empirical approaches to the study of language evolution. *Psychonomic bulletin & review*, 24(1), 3–33.

Gagliano, M. et al. (2017). Tuned in: plant roots use sound to locate water. *Oecologia*, *184*(1), 151–160.

Hamilton, L. et al. (2021). Parallel and distributed encoding of speech across human auditory cortex. *Cell*, *184*(18), 4626–4639.

Heesink, L. et al. (2017). Anger and aggression problems in veterans are associated with an increased acoustic startle reflex. *Biological Psychology*, *123*, 119–125.

Magrassi, L. et al. (2015). Sound representation in higher language areas during language generation. *Proceedings of the National Academy of Sciences*, *112*(6), 1868–1873.

McFadden, D. (1998). Sex differences in the auditory system. *Developmental Neuropsychology*, *14*(2–3), 261–298.

Mesgarani, N. et al. (2014). Phonetic feature encoding in human superior temporal gyrus. *Science*, *343*(6174), 1006–1010.

Pagel, M. (2017). What is human language, when did it evolve and why should we care?. *BMC Biology*, *15*(1), 1–6.

Sauter, D. et al. (2010). Cross-cultural recognition of basic emotions through nonverbal emotional vocalizations. *Proceedings of the National Academy of Sciences*, *107*(6), 2408–2412.

Schneider, D. & Mooney, R. (2018). How movement modulates hearing. *Annual Review of Neuroscience*, *41*, 553.

Shahin, A. et al. (2009). Neural mechanisms for illusory filling-in of degraded speech. *Neuroimage*, *44*(3), 1133–1143.

Vinnik, E. et al. (2011). Individual differences in sound-in-noise perception are related to the strength of short-latency neural responses to noise. *PloS One*, *6*(2), e17266.

Zatorre, R. & Salimpoor, V. (2013). From perception to pleasure: music and its neural substrates. *Proceedings of the National Academy of Sciences*, *110*, 10430–10437.

## Scents and Scents Ability

Aqrabawi, A. & Kim, J. (2020). Olfactory memory representations are stored in the anterior olfactory nucleus. *Nature Communications, 11*(1), 1–8.

Cameron, E. et al. (2016). The accuracy, consistency, and speed of odor and picture naming. *Chemosensory Perception, 9*(2), 69–78.

Chu, S. & Downes, J. (2000). Odour-evoked autobiographical memories. *Chemical Senses, 25*(1), 111–116.

Classen, C. (1992). The odor of the other: olfactory symbolism and cultural categories. *Ethos, 20*(2), 133–166.

Classen, C. (1999). Other ways to wisdom: Learning through the senses across cultures. *International Review of Education, 45*(3), 269–280.

Dahmani, L. et al. (2018). An intrinsic association between olfactory identification and spatial memory in humans. *Nature Communications, 9*(1), 1–12.

de Groot, J. et al. (2020). Encoding fear intensity in human sweat. *Philosophical Transactions of the Royal Society B, 375*(1800), 20190271.

de Wijk, R. & Zijlstra, S. (2012). Differential effects of exposure to ambient vanilla and citrus aromas on mood, arousal and food choice. *Flavour, 1*(1), 1–7.

Derti, A. et al. (2010). Absence of evidence for MHC–dependent mate selection within HapMap populations. *PLoS Genetics, 6*(4), e1000925.

Frumin, I. et al. (2014). Does a unique olfactory genome imply a unique olfactory world?. *Nature Neuroscience, 17*(1), 6–8.

Hackländer, R. et al. (2019). An in-depth review of the methods, findings, and theories associated with odor-evoked autobiographical memory. *Psychonomic Bulletin & Review, 26*(2), 401–429.

Havlicek, J., & Lenochova, P. (2006). The effect of meat consumption on body odor attractiveness. *Chemical Senses, 31*(8), 747–752.

Havlíček, J. et al. (2017). Individual variation in body odor. In *Springer Handbook of Odor*.

Herz, R. (2009). Aromatherapy facts and fictions. *International Journal of Neuroscience, 119*(2), 263–290.

Herz, R. & von Clef, J. (2001). The influence of verbal labeling on the perception of odors: evidence for olfactory illusions? *Perception, 30*(3), 381–391.

Jacobs, L. (2012). From chemotaxis to the cognitive map. *Proceedings of the National Academy of Sciences, 109*, 10693–10700.

# Further Reading and Selected References

Jacobs, L. et al. (2015). Olfactory orientation and navigation in humans. *PLoS One, 10*(6), e0129387.

Kontaris, I. et al. (2020). Behavioral and neurobiological convergence of odor, mood and emotion. *Frontiers in Behavioral Neuroscience, 14*, 35.

Laska, M. (2017). Human and animal olfactory capabilities compared. In *Springer Handbook of Odor*

Logan, D. (2014). Do you smell what I smell? Genetic variation in olfactory perception. *Biochemical Society Transactions, 42*(4), 861–865.

Majid, A. (2021). Human olfaction at the intersection of language, culture, and biology. *Trends in Cognitive Sciences, 25*(2), 111–123.

Majid, A., & Burenhult, N. (2014). Odors are expressible in language, as long as you speak the right language. *Cognition, 130*(2), 266–270.

McGann, J. P. (2017). Poor human olfaction is a 19th-century myth. *Science, 356*(6338), eaam7263.

Minhas, G. et al. (2018). Structural basis of malodour precursor transport in the human axilla. *Elife, 7*, e34995.

O'Mahony, M. (1978). Smell illusions and suggestion: Reports of smells contingent on tones played on television and radio. *Chemical Senses, 3*(2), 183–189.

Perl, O. et al. (2020). Are humans constantly but subconsciously smelling themselves?. *Philosophical Transactions of the Royal Society B, 375*(1800), 20190372.

Porter, J. et al. (2007). Mechanisms of scent-tracking in humans. *Nature neuroscience, 10*(1), 27–29.

Prokop-Prigge, K. et al. (2016). The effect of ethnicity on human axillary odorant production. *Journal of Chemical Ecology, 42*(1), 33–39.

Reicher, S. et al. (2016). Core disgust is attenuated by ingroup relations. *Proceedings of the National Academy of Sciences, 113*(10), 2631–2635.

Rimkute, J. et al. (2016). The effects of scent on consumer behaviour. *International Journal of Consumer Studies, 40*(1), 24–34.

Roberts, S. et al. (2008). MHC-correlated odour preferences in humans and the use of oral contraceptives. *Proceedings of the Royal Society B: Biological Sciences, 275*(1652), 2715–2722.

Roberts, S. et al. (2020). Human olfactory communication. *Philosophical Transactions of the Royal Society B, 375*(1800), 20190258.

Ross, A. et al. (2019). The skin microbiome of vertebrates. *Microbiome, 7*, 1–14.

Sarafoleanu, C. et al. (2009). The importance of the olfactory sense in the human behavior and evolution. *Journal of Medicine and Life*, 2(2), 196.

Shirasu, M., & Touhara, K. (2011). The scent of disease. *The Journal of Biochemistry*, 150(3), 257–266.

Sorokowska, A. et al. (2012). Does personality smell?. *European Journal of Personality*, 26(5), 496–503.

Sorokowska, A. et al. (2013). Olfaction and environment. *PloS One*, 8(7), e69203.

Sorokowski, P. et al. (2019). Sex differences in human olfaction: a meta-analysis. *Frontiers in Psychology*, 10, 242.

Spence, C. (2021). The scent of attraction and the smell of success: crossmodal influences on person perception. *Cognitive Research: Principles and Implications*, 6(1), 1–33.

Stancak, A. et al. (2015). Unpleasant odors increase aversion to monetary losses. *Biological psychology*, 107, 1–9.

Stevenson, R. & Repacholi, B. (2005). Does the source of an interpersonal odour affect disgust? *European Journal of Social Psychology*, 35(3), 375–401.

Trimmer, C. et al. (2019). Genetic variation across the human olfactory receptor repertoire alters odor perception. *Proceedings of the National Academy of Sciences*, 116(19), 9475–9480.

Übel, S. et al. (2017). Affective evaluation of one's own and others' body odor: the role of disgust proneness. *Perception*, 46(12), 1427–1433.

Villemure, C. et al. (2003). Effects of odors on pain perception. *Pain*, 106(1–2), 101–108.

Wedekind, C. et al. (1995). MHC-dependent mate preferences in humans. *Proceedings of the Royal Society of London B*, 260(1359), 245–249.

Wyatt, T. (2020). Reproducible research into human chemical communication by cues and pheromones: learning from psychology's renaissance. *Philosophical Transactions of the Royal Society B*, 375(1800), 20190262.

Zhang, S., & Manahan-Vaughan, D. (2015). Spatial olfactory learning contributes to place field formation in the hippocampus. *Cerebral Cortex*, 25(2), 423–432.

## Accounting for Taste

Armitage, R. et al. (2021). Understanding sweet-liking phenotypes and their implications for obesity. *Physiology & Behavior*, 235, 113398.

# Further Reading and Selected References

Asarian, L., & Geary, N. (2013). Sex differences in the physiology of eating. *American Journal of Physiology-Regulatory, Integrative and Comparative Physiology, 305*(11), R1215-R1267.

Bakke, A. et al. (2018). Mary Poppins was right: Adding small amounts of sugar or salt reduces the bitterness of vegetables. *Appetite, 126,* 90–101.

Behrens, M., & Meyerhof, W. (2011). Gustatory and extragustatory functions of mammalian taste receptors. *Physiology & Behavior, 105*(1), 4–13.

Benson, P. et al. (2012). Bitter taster status predicts susceptibility to vection-induced motion sickness and nausea. *Neurogastroenterology & Motility, 24*(2), 134-e86.

Breslin, P. (1996). Interactions among salty, sour and bitter compounds. *Trends in Food Science & Technology, 7*(12), 390–399.

Breslin, P. (2013). An evolutionary perspective on food and human taste. *Current Biology, 23*(9), R409-R418.

Briand, L., & Salles, C. (2016). Taste perception and integration. In *Flavor.* Woodhead Publishing.

Costanzo, A. et al. (2019). A low-fat diet up-regulates expression of fatty acid taste receptor gene FFAR4 in fungiform papillae in humans. *British Journal of Nutrition, 122*(11), 1212–1220.

Dalton, P. et al. (2000). The merging of the senses: integration of subthreshold taste and smell. *Nature Neuroscience, 3*(5), 431–432.

Doty, R. (2015). *Handbook of Olfaction and Gustation.* John Wiley & Sons.

Eisenstein, M. (2010). Taste: More than meets the mouth. *Nature, 468*(7327), S18-S19.

Forestell, C. (2017). Flavor perception and preference development in human infants. *Annals of Nutrition and Metabolism, 70,* 17–25.

Green, B. & George, P. (2004). 'Thermal taste' predicts higher responsiveness to chemical taste and flavor. *Chemical Senses, 29*(7), 617–628.

Hummel, T. et al. (2006). Perceptual differences between chemical stimuli presented through the ortho – or retronasal route. *Flavour and Fragrance Journal, 21*(1), 42–47.

Karagiannaki, K. et al. (2021). Determining optimal exposure frequency for introducing a novel vegetable among children. *Foods, 10*(5), 913.

Keast, R. et al. (2021). Macronutrient sensing in the oral cavity and gastrointestinal tract: alimentary tastes. *Nutrients, 13*(2), 667.

Lenfant, F. et al. (2013). Impact of the shape on sensory properties of individual dark chocolate pieces. *LWT-Food Science and Technology, 51*(2), 545–552.

Martin, L. & Sollars, S. (2017). Contributory role of sex differences in the variations of gustatory function. *Journal of Neuroscience Research*, 95(1–2), 594–603.

Maruyama, Y. et al. (2012). Kokumi substances, enhancers of basic tastes, induce responses in calcium-sensing receptor expressing taste cells. *PLoS One*, 7(4), e34489.

Reed, D. & Knaapila, A. (2010). Genetics of taste and smell: poisons and pleasures. *Progress in Molecular Biology*, 94, 213–240.

Shizukuda, S. et al. (2018). Influences of weight, age, gender, genetics, diseases, and ethnicity on bitterness perception. *Nutrire*, 43(1), 1–9.

Slack, J. (2016). Molecular pharmacology of chemesthesis. In *Chemosensory Transduction*. Academic Press.

Small, D. et al. (2005). Differential neural responses evoked by orthonasal versus retronasal odorant perception in humans. *Neuron*, 47(4), 593–605.

Spence, C. (2013). Multisensory flavour perception. *Current Biology*, 23(9), R365-R369.

Spence, C. (2015). Just how much of what we taste derives from the sense of smell? *Flavour*, 4(1), 1–10.

Spence, C., & Wang, Q. (2015). Wine and music (II): can you taste the music? Modulating the experience of wine through music and sound. *Flavour*, 4(1), 1–14.

Spence, C. et al. (2016). Eating with our eyes: From visual hunger to digital satiation. *Brain and cognition*, 110, 53–63.

Stevenson, R. et al. (2011). The role of taste and oral somatosensation in olfactory localization. *Quarterly Journal of Experimental Psychology*, 64(2), 224–240.

Wang, Y. et al. (2019). Metal ions activate the human taste receptor TAS2R7. *Chemical senses*, 44(5), 339–347.

Williams, J. et al. (2016). Exploring ethnic differences in taste perception. *Chemical senses*, 41(5), 449–456.

Yang, Q. et al. (2020). Exploring the relationships between taste phenotypes, genotypes, ethnicity, gender and taste perception. *Food Quality and Preference*, 83, 103928.

Yarmolinsky, D. et al. (2009). Common sense about taste: from mammals to insects. *Cell*, 139(2), 234–244.

Yohe, L. & Brand, P. (2018). Evolutionary ecology of chemosensation and its role in sensory drive. *Current Zoology*, 64(4), 525–533.

# Further Reading and Selected References

## Skin Sense

Ackerman, J. et al. (2010). Incidental haptic sensations influence social judgments and decisions. *Science*, 328(5986), 1712–1715.

Ardiel, E. & Rankin, C. (2010). The importance of touch in development. *Paediatrics & Child Health*, 15(3), 153–156.

Bartley, E. & Fillingim, R. (2013). Sex differences in pain: a brief review of clinical and experimental findings. *British Journal of Anaesthesia*, 111(1), 52–58.

Carpenter, C. et al. (2018). Human ability to discriminate surface chemistry by touch. *Materials Horizons*, 5(1), 70–77.

Coan, J. et al. (2006). Lending a hand: Social regulation of the neural response to threat. *Psychological Science*, 17(12), 1032–1039.

Corniani, G., & Saal, H. (2020). Tactile innervation densities across the whole body. *Journal of Neurophysiology*, 124(4), 1229–1240.

Craft, R. (2007). Modulation of pain by estrogens. *Pain*, 132, S3-S12.

Dubin, A. & Patapoutian, A. (2010). Nociceptors: the sensors of the pain pathway. *The Journal of Clinical Investigation*, 120(11), 3760–3772.

Feldman, R. et al. (2014). Maternal-preterm skin-to-skin contact enhances child physiologic organization and cognitive control across the first 10 years of life. *Biological Psychiatry*, 75(1), 56–64.

Field, T. (2010). Touch for socioemotional and physical well-being. *Developmental Review*, 30(4), 367–383.

Gallace, A., & Spence, C. (2010). The science of interpersonal touch. *Neuroscience & Biobehavioral Reviews*, 34(2), 246–259.

Gibson, J. (1933). Adaptation, after-effect and contrast in the perception of curved lines. *Journal of Experimental Psychology*, 16(1), 1.

Gilam, G. et al. (2020). What is the relationship between pain and emotion? *Neuron*, 107(1), 17–21.

Gindrat, A. et al. (2015). Use-dependent cortical processing from fingertips in touchscreen phone users. *Current Biology*, 25(1), 109–116.

Goldstein, P. et al. (2018). Brain-to-brain coupling during handholding is associated with pain reduction. *Proceedings of the National Academy of Sciences*, 115(11), E2528-E2537.

Guéguen, N., & Jacob, C. (2005). The effect of touch on tipping: an evaluation in a French bar. *International Journal of Hospitality Management*, 24(2), 295–299.

Hertenstein, M. et al. (2009). The communication of emotion via touch. *Emotion*, 9(4), 566.

Kelley, N. & Schmeichel, B. (2014). The effects of negative emotions on sensory perception. *Frontiers in Psychology*, *5*, 942.

Kraus, M. et al. (2010). Tactile communication, cooperation, and performance: an ethological study of the NBA. *Emotion*, *10*(5), 745.

Kung, C. (2005). A possible unifying principle for mechanosensation. *Nature*, *436*(7051), 647–654.

Mancini, F., Bauleo, A., Cole, J., Lui, F., Porro, C. A., Haggard, P., & Iannetti, G. D. (2014). Whole-body mapping of spatial acuity for pain and touch. *Annals of neurology*, *75*(6), 917–924.

McGlone, F., Wessberg, J., & Olausson, H. (2014). Discriminative and affective touch: sensing and feeling. *Neuron*, *82*(4), 737–755.

Orban, G. A., & Caruana, F. (2014). The neural basis of human tool use. *Frontiers in psychology*, *5*, 310.

Pawling, R. et al. (2017). C-tactile afferent stimulating touch carries a positive affective value. *PloS One*, *12*(3), e0173457.

Skedung, L. et al. (2013). Feeling small: exploring the tactile perception limits. *Scientific reports*, *3*(1), 1–6.

von Mohr, M. et al. (2017). The soothing function of touch: affective touch reduces feelings of social exclusion. *Scientific Reports*, *7*(1), 1–9.

Voss, P. (2011). Superior tactile abilities in the blind: is blindness required? *Journal of Neuroscience*, *31*(33), 11745–11747.

## The Kitchen Drawer of the Senses

Baiano, C. et al. (2021). Interactions between interoception and perspective-taking. *Neuroscience & Biobehavioral Reviews*, *130*, 252–262.

Craig, A. (2003). Interoception: the sense of the physiological condition of the body. *Current Opinion in Neurobiology*, *13*(4), 500–505.

Fuchs, D. (2018). Dancing with gravity – Why the sense of balance is (the) fundamental. *Behavioral Sciences*, *8*(1), 7.

Garfinkel, S. et al. (2015). Knowing your own heart: distinguishing interoceptive accuracy from interoceptive awareness. *Biological psychology*, *104*, 65–74.

Holland, R. et al. (2008). Bats use magnetite to detect the earth's magnetic field. *PLoS One*, *3*(2), e1676.

Koeppel, C. et al. (2020). Interoceptive accuracy and its impact on neuronal responses to olfactory stimulation in the insular cortex. *Human Brain Mapping*, *41*(11), 2898–2908.

Paulus, M. & Stein, M. (2010). Interoception in anxiety and depression. *Brain structure and Function*, *214*(5), 451–463.

Sato, J. (2003). Weather change and pain. *International Journal of Biometeorology*, 47(2), 55–61.

Sato, J. et al. (2019). Lowering barometric pressure induces neuronal activation in the superior vestibular nucleus in mice. *PLoS One*, 14(1), e0211297.

Seth, A. & Friston, K. (2016). Active interoceptive inference and the emotional brain. *Philosophical Transactions of the Royal Society B*, 371(1708), 20160007.

Smith, R. et al. (2021). Perceptual insensitivity to the modulation of interoceptive signals in depression, anxiety, and substance use disorders. *Scientific Reports*, 11(1), 1–14.

Suzuki, K. et al. (2013). Multisensory integration across exteroceptive and interoceptive domains modulates self-experience in the rubber-hand illusion. *Neuropsychologia*, 51(13), 2909–2917.

Wang, C. et al. (2019). Transduction of the geomagnetic field as evidenced from alpha-band activity in the human brain. *eNeuro*.

Xu, J. et al. (2021). Magnetic sensitivity of cryptochrome 4 from a migratory songbird. *Nature*, 594(7864), 535–540.

## The Weave of Perception

Albright, T. (2017). Why eyewitnesses fail. *Proceedings of the National Academy of Sciences*, 114(30), 7758–7764.

Brang, D., & Ramachandran, V. (2011). Why do people hear colors and taste words?. *PLoS biology*, 9(11), e1001205.

Cecere, R. et al. (2015). Individual differences in alpha frequency drive crossmodal illusory perception. *Current Biology*, 25(2), 231–235.

Dematte, M. et al.. (2006). Cross-modal interactions between olfaction and touch. *Chemical Senses*, 31(4), 291–300.

Ernst, M. & Banks, M. (2002). Humans integrate visual and haptic information in a statistically optimal fashion. *Nature*, 415(6870), 429–433.

Ernst, M. & Bülthoff, H. (2004). Merging the senses into a robust percept. *Trends in Cognitive Sciences*, 8(4), 162–169.

Gau, R. et al. (2020). Resolving multisensory and attentional influences across cortical depth in sensory cortices. *Elife*, 9.

Hadley, L. et al. (2019). Speech, movement, and gaze behaviours during dyadic conversation in noise. *Scientific Reports*, 9(1), 1–8.

Jadauji, J. et al. (2012). Modulation of olfactory perception by visual cortex stimulation. *Journal of Neuroscience*, 32(9), 3095–3100.

Majid, A. et al. (2018). Differential coding of perception in the world's languages. *Proceedings of the National Academy of Sciences*, *115*(45), 11369–11376.

O'Callaghan, C. (2017). Synesthesia vs. Crossmodal.*Sensory blending*.

Rigato, S. et al. (2016). Multisensory signalling enhances pupil dilation. *Scientific Reports*, *6*(1), 1–9.

Schifferstein, H. & Spence, C. (2008). Multisensory product experience. In *Product Experience*. Elsevier.

Spence, C. (2011). Crossmodal correspondences. *Attention, Perception, & Psychophysics*, *73*(4), 971–995.

Teichert, M., & Bolz, J. (2018). How senses work together: cross-modal interactions between primary sensory cortices. *Neural plasticity*, *2018*.

Theeuwes, J. et al. (2007). Cross-modal interactions between sensory modalities: Implications for the design of multisensory displays. *Attention: From theory to practice*, 196–205.

Van Den Brink, R. et al. (2016). Pupil diameter tracks lapses of attention. *PLoS One*, *11*(10), e0165274.

Van Leeuwen, T. et al. (2015). The merit of synesthesia for consciousness research. *Frontiers in Psychology*, *6*, 1850.

Wise, R. et al. (2014). An examination of the causes and solutions to eyewitness error. *Frontiers in Psychiatry*, *5*, 102.

# Acknowledgments

I've been incredibly fortunate to have the support of so many people in the writing of this book. My agent, Max Edwards, has been indefatigable with his insight and encouragement. Then there's a whole team of wonderful people at Profile whose input has been absolutely invaluable. Particular mention must go to Nick Humphrey, who did an outstanding job in knocking the manuscript into shape, Emily Frisella, who has been a peerless guide throughout the process, and Alex Elam, whose efforts in spreading the word around the world have astonished me! I'd also like to thank Fran Fabriczki for her supremely helpful comments on the copy edit. Over the other side of the Atlantic, Emma Berry and her team at Basic Books have been a great source of support and wisdom throughout. I've loved working with you all, and I appreciate everything you've done.

Then there are so many people who I've had the privilege to work alongside, who've each contributed in their unique ways. Stella Encel not only read a filthy first draft of this book, but made some extremely valuable comments, and engaged me in discussions that fundamentally shaped my thoughts. Past members of my research group, including Alex Wilson, James 'Teddy' Herbert-Read, Alicia Burns, Matt Hansen, Mia Kent and Chris Reid, have each inspired me to think in new, and better, ways. Callum Steven, part-time bat and the poster child for Dunning-Kruger, has my endless gratitude for his advice and support. I'd also like to thank Rikki Jodelko and Hattie Jodelka for their encouragement, which meant (and means) a great deal to me.

# Index

# Index

# Index

hominids 73
honeybee 45, 87
Humboldt, Alexander von 16
Hume, David 54
hyperosmics 120–1

## I

Icarus 228
Ig Nobel Prize 233
Ikeda, Kikunae 153
immune system 114, 124, 164, 195, 224
inattention blindness 16–17
Inbar, Yoel 137
indole 120
infrared 42–3
infrasonic range 91
inner ear 87, 88–90, 215, 231, 237, 238
inner hair cells 99–100
insular cortex 64–5, 196–7
internal model 7–8, 11
International Space Station 228
interoceptive senses 241, 244, 245–6, 247
intrusive sounds 63–6
invertebrates 85–7, 231
iPhone 28, 216, 277
iris 24, 25
iron 234

## J

jacaranda 1–2
Jacobs, Lucia 142
Jadauji, Jahan 259
Jahai people 119
*Jaws* (film) 60
Jay, Stacey 147
jazz 79, 82

jewel beetle 268–9
Jobs, Steve 49
Jodelko, Rikki 69
Jonas, H.: 'The Nobility of Sight' 15
Jourard, Sidney 223
just noticeable difference 92–3

## K

Kaiser, Kurt 212
Kālidāsa 132
Kangaroo Care 193
Kant, Immanuel 54
Kay, Paul: *Basic Color Terms: Their Universality and Evolution* 35
Keller, Helen 219
Kidd, William 133
King, John 175
kissing 124
Köhler, Wolfgang 256
kokumi 157–8
Kölliker, Albert von 27
kombu 153
Korean language 37
Krakatoa, eruption of, Indonesia (1883) 95–6
Kraus, Michael 196

## L

Landells, George 175
language
    brain shaping and 97
    hearing and 71–7, 96–9
    linguistic relativism 34
    McGurk effect and 266
    smell and 116, 117, 119, 130, 265
    taste and 265, 267
    touch and 189, 191, 198
    vision and 34–8, 41, 265

# Index

**Ashley Ward** is a professor and director of the Animal Behaviour Lab at the University of Sydney, where he researches social behavior, learning, and communication in animals. His work has been published in top journals including *PNAS*, *Biological Reviews*, and *Current Biology*. The author of *The Social Lives of Animals*, he lives in Sydney, Australia.